FREE Test Taking Tips DVD Offer

To help us better serve you, we have developed a Test Taking Tips DVD that we would like to give you for FREE. **This DVD covers world-class test taking tips that you can use to be even more successful when you are taking your test.**

All that we ask is that you email us your feedback about your study guide. Please let us know what you thought about it – whether that is good, bad or indifferent.

To get your **FREE Test Taking Tips DVD**, email freedvd@studyguideteam.com with "FREE DVD" in the subject line and the following information in the body of the email:

> a. The title of your study guide.
>
> b. Your product rating on a scale of 1-5, with 5 being the highest rating.
>
> c. Your feedback about the study guide. What did you think of it?
>
> d. Your full name and shipping address to send your free DVD.

If you have any questions or concerns, please don't hesitate to contact us at freedvd@studyguideteam.com.

Thanks again!

FSA Practice Grade 3 Math Test Prep for the Florida Standards Assessment [3rd Edition Book]

Joshua Rueda

Interested in buying more than 10 copies of our product? Contact us about bulk discounts:
bulkorders@studyguideteam.com

ISBN 13: 9781637756546
ISBN 10: 1637756542

Table of Contents

Quick Overview

As you draw closer to taking your exam, effective preparation becomes more and more important. Thankfully, you have this study guide to help you get ready. Use this guide to help keep your studying on track and refer to it often.

This study guide contains several key sections that will help you be successful on your exam. The guide contains tips for what you should do the night before and the day of the test. Also included are test-taking tips. Knowing the right information is not always enough. Many well-prepared test takers struggle with exams. These tips will help equip you to accurately read, assess, and answer test questions.

A large part of the guide is devoted to showing you what content to expect on the exam and to helping you better understand that content. In this guide are practice test questions so that you can see how well you have grasped the content. Then, answer explanations are provided so that you can understand why you missed certain questions.

Don't try to cram the night before you take your exam. This is not a wise strategy for a few reasons. First, your retention of the information will be low. Your time would be better used by reviewing information you already know rather than trying to learn a lot of new information. Second, you will likely become stressed as you try to gain a large amount of knowledge in a short amount of time. Third, you will be depriving yourself of sleep. So be sure to go to bed at a reasonable time the night before. Being well-rested helps you focus and remain calm.

Be sure to eat a substantial breakfast the morning of the exam. If you are taking the exam in the afternoon, be sure to have a good lunch as well. Being hungry is distracting and can make it difficult to focus. You have hopefully spent lots of time preparing for the exam. Don't let an empty stomach get in the way of success!

When travelling to the testing center, leave earlier than needed. That way, you have a buffer in case you experience any delays. This will help you remain calm and will keep you from missing your appointment time at the testing center.

Be sure to pace yourself during the exam. Don't try to rush through the exam. There is no need to risk performing poorly on the exam just so you can leave the testing center early. Allow yourself to use all of the allotted time if needed.

Remain positive while taking the exam even if you feel like you are performing poorly. Thinking about the content you should have mastered will not help you perform better on the exam.

Once the exam is complete, take some time to relax. Even if you feel that you need to take the exam again, you will be well served by some down time before you begin studying again. It's often easier to convince yourself to study if you know that it will come with a reward!

Test-Taking Strategies

1. Predicting the Answer

When you feel confident in your preparation for a multiple-choice test, try predicting the answer before reading the answer choices. This is especially useful on questions that test objective factual knowledge. By predicting the answer before reading the available choices, you eliminate the possibility that you will be distracted or led astray by an incorrect answer choice. You will feel more confident in your selection if you read the question, predict the answer, and then find your prediction among the answer choices. After using this strategy, be sure to still read all of the answer choices carefully and completely. If you feel unprepared, you should not attempt to predict the answers. This would be a waste of time and an opportunity for your mind to wander in the wrong direction.

2. Reading the Whole Question

Too often, test takers scan a multiple-choice question, recognize a few familiar words, and immediately jump to the answer choices. Test authors are aware of this common impatience, and they will sometimes prey upon it. For instance, a test author might subtly turn the question into a negative, or he or she might redirect the focus of the question right at the end. The only way to avoid falling into these traps is to read the entirety of the question carefully before reading the answer choices.

3. Looking for Wrong Answers

Long and complicated multiple-choice questions can be intimidating. One way to simplify a difficult multiple-choice question is to eliminate all of the answer choices that are clearly wrong. In most sets of answers, there will be at least one selection that can be dismissed right away. If the test is administered on paper, the test taker could draw a line through it to indicate that it may be ignored; otherwise, the test taker will have to perform this operation mentally or on scratch paper. In either case, once the obviously incorrect answers have been eliminated, the remaining choices may be considered. Sometimes identifying the clearly wrong answers will give the test taker some information about the correct answer. For instance, if one of the remaining answer choices is a direct opposite of one of the eliminated answer choices, it may well be the correct answer. The opposite of obviously wrong is obviously right! Of course, this is not always the case. Some answers are obviously incorrect simply because they are irrelevant to the question being asked. Still, identifying and eliminating some incorrect answer choices is a good way to simplify a multiple-choice question.

4. Don't Overanalyze

Anxious test takers often overanalyze questions. When you are nervous, your brain will often run wild, causing you to make associations and discover clues that don't actually exist. If you feel that this may be a problem for you, do whatever you can to slow down during the test. Try taking a deep breath or counting to ten. As you read and consider the question, restrict yourself to the particular words used by the author. Avoid thought tangents about what the author *really* meant, or what he or she was *trying* to say. The only things that matter on a multiple-choice test are the words that are actually in the question. You must avoid reading too much into a multiple-choice question, or supposing that the writer meant something other than what he or she wrote.

5. No Need for Panic

It is wise to learn as many strategies as possible before taking a multiple-choice test, but it is likely that you will come across a few questions for which you simply don't know the answer. In this situation, avoid panicking. Because most multiple-choice tests include dozens of questions, the relative value of a single wrong answer is small. As much as possible, you should compartmentalize each question on a multiple-choice test. In other words, you should not allow your feelings about one question to affect your success on the others. When you find a question that you either don't understand or don't know how to answer, just take a deep breath and do your best. Read the entire question slowly and carefully. Try rephrasing the question a couple of different ways. Then, read all of the answer choices carefully. After eliminating obviously wrong answers, make a selection and move on to the next question.

6. Confusing Answer Choices

When working on a difficult multiple-choice question, there may be a tendency to focus on the answer choices that are the easiest to understand. Many people, whether consciously or not, gravitate to the answer choices that require the least concentration, knowledge, and memory. This is a mistake. When you come across an answer choice that is confusing, you should give it extra attention. A question might be confusing because you do not know the subject matter to which it refers. If this is the case, don't eliminate the answer before you have affirmatively settled on another. When you come across an answer choice of this type, set it aside as you look at the remaining choices. If you can confidently assert that one of the other choices is correct, you can leave the confusing answer aside. Otherwise, you will need to take a moment to try to better understand the confusing answer choice. Rephrasing is one way to tease out the sense of a confusing answer choice.

7. Your First Instinct

Many people struggle with multiple-choice tests because they overthink the questions. If you have studied sufficiently for the test, you should be prepared to trust your first instinct once you have carefully and completely read the question and all of the answer choices. There is a great deal of research suggesting that the mind can come to the correct conclusion very quickly once it has obtained all of the relevant information. At times, it may seem to you as if your intuition is working faster even than your reasoning mind. This may in fact be true. The knowledge you obtain while studying may be retrieved from your subconscious before you have a chance to work out the associations that support it. Verify your instinct by working out the reasons that it should be trusted.

8. Key Words

Many test takers struggle with multiple-choice questions because they have poor reading comprehension skills. Quickly reading and understanding a multiple-choice question requires a mixture of skill and experience. To help with this, try jotting down a few key words and phrases on a piece of scrap paper. Doing this concentrates the process of reading and forces the mind to weigh the relative importance of the question's parts. In selecting words and phrases to write down, the test taker thinks about the question more deeply and carefully. This is especially true for multiple-choice questions that are preceded by a long prompt.

9. Subtle Negatives

One of the oldest tricks in the multiple-choice test writer's book is to subtly reverse the meaning of a question with a word like *not* or *except*. If you are not paying attention to each word in the question, you can easily be led astray by this trick. For instance, a common question format is, "Which of the following is...?" Obviously, if the question instead is, "Which of the following is not...?," then the answer will be quite different. Even worse, the test makers are aware of the potential for this mistake and will include one answer choice that would be correct if the question were not negated or reversed. A test taker who misses the reversal will find what he or she believes to be a correct answer and will be so confident that he or she will fail to reread the question and discover the original error. The only way to avoid this is to practice a wide variety of multiple-choice questions and to pay close attention to each and every word.

10. Reading Every Answer Choice

It may seem obvious, but you should always read every one of the answer choices! Too many test takers fall into the habit of scanning the question and assuming that they understand the question because they recognize a few key words. From there, they pick the first answer choice that answers the question they believe they have read. Test takers who read all of the answer choices might discover that one of the latter answer choices is actually *more* correct. Moreover, reading all of the answer choices can remind you of facts related to the question that can help you arrive at the correct answer. Sometimes, a misstatement or incorrect detail in one of the latter answer choices will trigger your memory of the subject and will enable you to find the right answer. Failing to read all of the answer choices is like not reading all of the items on a restaurant menu: you might miss out on the perfect choice.

11. Spot the Hedges

One of the keys to success on multiple-choice tests is paying close attention to every word. This is never truer than with words like almost, most, some, and sometimes. These words are called "hedges" because they indicate that a statement is not totally true or not true in every place and time. An absolute statement will contain no hedges, but in many subjects, the answers are not always straightforward or absolute. There are always exceptions to the rules in these subjects. For this reason, you should favor those multiple-choice questions that contain hedging language. The presence of qualifying words indicates that the author is taking special care with his or her words, which is certainly important when composing the right answer. After all, there are many ways to be wrong, but there is only one way to be right! For this reason, it is wise to avoid answers that are absolute when taking a multiple-choice test. An absolute answer is one that says things are either all one way or all another. They often include words like *every*, *always*, *best*, and *never*. If you are taking a multiple-choice test in a subject that doesn't lend itself to absolute answers, be on your guard if you see any of these words.

12. Long Answers

In many subject areas, the answers are not simple. As already mentioned, the right answer often requires hedges. Another common feature of the answers to a complex or subjective question are qualifying clauses, which are groups of words that subtly modify the meaning of the sentence. If the question or answer choice describes a rule to which there are exceptions or the subject matter is complicated, ambiguous, or confusing, the correct answer will require many words in order to be expressed clearly and accurately. In essence, you should not be deterred by answer choices that seem excessively long. Oftentimes, the author of the text will not be able to write the correct answer without

offering some qualifications and modifications. Your job is to read the answer choices thoroughly and completely and to select the one that most accurately and precisely answers the question.

13. Restating to Understand

Sometimes, a question on a multiple-choice test is difficult not because of what it asks but because of how it is written. If this is the case, restate the question or answer choice in different words. This process serves a couple of important purposes. First, it forces you to concentrate on the core of the question. In order to rephrase the question accurately, you have to understand it well. Rephrasing the question will concentrate your mind on the key words and ideas. Second, it will present the information to your mind in a fresh way. This process may trigger your memory and render some useful scrap of information picked up while studying.

14. True Statements

Sometimes an answer choice will be true in itself, but it does not answer the question. This is one of the main reasons why it is essential to read the question carefully and completely before proceeding to the answer choices. Too often, test takers skip ahead to the answer choices and look for true statements. Having found one of these, they are content to select it without reference to the question above. Obviously, this provides an easy way for test makers to play tricks. The savvy test taker will always read the entire question before turning to the answer choices. Then, having settled on a correct answer choice, he or she will refer to the original question and ensure that the selected answer is relevant. The mistake of choosing a correct-but-irrelevant answer choice is especially common on questions related to specific pieces of objective knowledge. A prepared test taker will have a wealth of factual knowledge at his or her disposal, and should not be careless in its application.

15. No Patterns

One of the more dangerous ideas that circulates about multiple-choice tests is that the correct answers tend to fall into patterns. These erroneous ideas range from a belief that B and C are the most common right answers, to the idea that an unprepared test-taker should answer "A-B-A-C-A-D-A-B-A." It cannot be emphasized enough that pattern-seeking of this type is exactly the WRONG way to approach a multiple-choice test. To begin with, it is highly unlikely that the test maker will plot the correct answers according to some predetermined pattern. The questions are scrambled and delivered in a random order. Furthermore, even if the test maker was following a pattern in the assignation of correct answers, there is no reason why the test taker would know which pattern he or she was using. Any attempt to discern a pattern in the answer choices is a waste of time and a distraction from the real work of taking the test. A test taker would be much better served by extra preparation before the test than by reliance on a pattern in the answers.

FREE DVD OFFER

Don't forget that doing well on your exam includes both understanding the test content and understanding how to use what you know to do well on the test. We offer a completely FREE Test Taking Tips DVD that covers world class test taking tips that you can use to be even more successful when you are taking your test.

All that we ask is that you email us your feedback about your study guide. To get your **FREE Test Taking Tips DVD**, email freedvd@studyguideteam.com with "FREE DVD" in the subject line and the following information in the body of the email:

- The title of your study guide.
- Your product rating on a scale of 1-5, with 5 being the highest rating.
- Your feedback about the study guide. What did you think of it?
- Your full name and shipping address to send your free DVD.

Introduction to the FSA Grade 3 Mathematics Exam

Function of the Test

The Florida Standards Assessments (FSA) Grade 3 Mathematics exam is intended to assess third graders' mastery and achievement of Florida math standards. It is part of the compendium of FSA exams, which students take from grade 3 through grade 8 in the subject of math and from grade 3 through grade 10 in the subject of ELA. Tests are administered near the end of the academic year, as they are designed to evaluate the knowledge and skills obtained over the student's current academic grade level.

Test Administration

The FSA Grade 3 Mathematics test is typically given to third grade students at school during the first two weeks of May in the on specific state-determined dates. It is administered during two days, with an 80-minute session each day. It is only offered in a paper-based format.

Accommodations are available for students with documented disabilities or who are English Language Learners if they have an IEP or Section 504 plan. Examples of available accommodations include large-print text booklets and booklets printed in braille. Calculators are not allowed on the exam.

Test Format

The test is conducted in two 80-minute sessions for a total test time of 160 minutes. There are between 60-64 questions on the test. The test items are designed to measure achievement of the grade 3 math standards set forth by the state of Florida. Questions may be of a variety of formats including multiple choice, multiselect, Graphic Response Item Display (GRID), and an equation editor format where students can create a response using a toolbar with mathematical symbols. The following table details the category and percentage of questions encountered on the test as well as the number of standards pertaining to each category:

Category	Percentage of Questions	# of Standards Assessed
Operations, Algebraic Thinking, and Numbers in Base Ten	48%	12
Numbers and Operations – Fractions	17%	4
Measurement, Data, and Geometry	35%	11

Scoring

Student performance on the Grade 3 Mathematics exam is categorized into five achievement levels. Level 1 achievement, which correlates to a scaled score of 240-284, indicates that a student's mastery of the standards is "inadequate" and that they are highly likely to need a substantial amount of support in the fourth grade. Level 2 achievement, which correlates to a scaled score of 285-296, is deemed "below satisfactory," and indicates that the student is likely to need a substantial amount of support in the fourth grade. Level 3 achievement, which correlates to a scaled score of 297-310, is considered "satisfactory," and indicates that the student may need support in the fourth grade. Achieving Level 4, which correlates to a scaled score of 311-326, is considered "proficient," and shows the student is likely to excel in the fourth grade. Lastly, Level 5 achievement, which correlates to a scaled score of 327-360, is considered "mastery" and indicative of a student who is highly likely to excel in the fourth grade. The passing score is 297, which is the lowest score in the Level 3 score range.

Operations and Algebraic Thinking

Representing and Solving Problems Involving Multiplication and Division

Interpreting Products of Whole Numbers

Multiplication involves taking multiple copies of one number. The sign designating a multiplication operation is the x symbol. The result is called the **product**. For example, $9 \times 6 = 54$. Multiplication means adding together one number in the equation as many times as the number on the other side of the equation:

$$9 \times 6 = 9 + 9 + 9 + 9 + 9 + 9 = 54$$

$$9 \times 6 = 6 + 6 + 6 + 6 + 6 + 6 + 6 + 6 + 6 = 54$$

The product, 54, of 9×6 can be thought of as six groups of nine, or nine groups of six.

Consider the problem 3×4. A model can be used to visualize these groups.

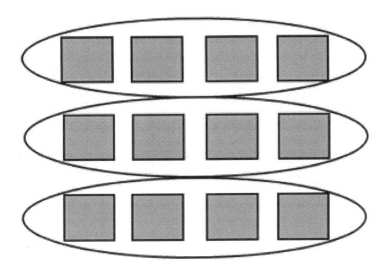

In this first graphic (above), there are three groups of four squares. The product of 3×4 is 12 because there are 12 total boxes. The same problem is displayed in the alternative grouping below:

Here there are still 12 boxes that model 3×4, but there are four groups with three boxes per group.

Interpreting Whole-Number Quotients of Whole Numbers as Partitioning Objects into Equal Shares

Division problems involve a total amount, a number of groups having the same amount, and a number of items within each group. The difference between multiplication and division is that the starting point is the total amount. It then gets divided into equal amounts.

For example, the equation is $18 \div 6 = 3$.

18 is the total number of items, which is being divided into 6 different groups. In order to do so, 3 items go into each group. Also, 6 and 3 are interchangeable. So, the 18 items could be divided into 3 groups of 6 items each. Therefore, different types of word problems can arise from this equation. For example, here are three types of problems:

- A boy needs 48 pieces of chalk. If there are 8 pieces in each box, how many boxes should he buy?

- A boy has 48 pieces of chalk. If each box has 6 pieces in it, how many boxes did he buy?

- A boy has partitioned all of his chalk into 8 piles, with 6 pieces in each pile. How many pieces does he have in total?

Each one of these questions involves the same equation. The third question can easily utilize the multiplication equation $8 \times 6 = ?$ instead of division. The answers are 6, 8, and 48.

Division can also be displayed through arrays. Using the previous jewelry store example, work backward. If the total number of sales is 12, how many sales occur evenly over 3 days? Or how many sales occur evenly over 4 days? The first question can be seen as follows:

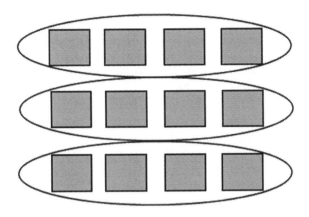

This demonstrates $12 \div 3 = 4$.

The answer for the second question looks like this:

This demonstrates $12 \div 4 = 3$.

Using Multiplication and Division Within 100 to Solve Word Problems

There are a variety of real-world situations in which multiplication and division are used to solve problems. The tables below display some of the most common scenarios.

	Unknown Product	**Unknown Group Size**	**Unknown Number of Groups**
Equal groups	There are 5 students, and each student has 4 pieces of candy. How many pieces of candy are there in all? $5 \times 4 = ?$	14 pieces of candy are shared equally by 7 students. How many pieces of candy does each student have? $7 \times ? = 14$ Solved by inverse operations $14 \div 7 = ?$	If 18 pieces of candy are to be given out 3 to each student, how many students will get candy? $? \times 3 = 18$ Solved by inverse operations $18 \div 3 = ?$

	Unknown Product	**Unknown Factor**	**Unknown Factor**
Arrays	There are 5 rows of students with 3 students in each row. How many students are there? $5 \times 3 = ?$	If 16 students are arranged into 4 equal rows, how many students will be in each row? $4 \times ? = 16$ Solved by inverse operations $16 \div 4 = ?$	If 24 students are arranged into an array with 6 columns, how many rows are there? $? \times 6 = 24$ Solved by inverse operations $24 \div 6 = ?$

	Larger Unknown	**Smaller Unknown**	**Multiplier Unknown**
Comparing	A small popcorn costs $1.50. A large popcorn costs 3 times as much as a small popcorn. How much does a large popcorn cost? $1.50 \times 3 = ?$	A large soda costs $6 and that is 2 times as much as a small soda costs. How much does a small soda cost? $2 \times ? = 6$ Solved by inverse operations $6 \div 2 = ?$	A large pretzel costs $3 and a small pretzel costs $2. How many times as much does the large pretzel cost as the small pretzel? $? \times 2 = 3$ Solved by inverse operations $3 \div 2 = ?$

Concrete objects can also be used to solve one- and two-step problems involving multiplication and division. Tools such as tiles, blocks, beads, and hundred charts are used to model problems. For example, a hundred chart (10×10) and beads can be used to model multiplication. If multiplying 5 by 4, beads are placed across 5 rows and down 4 columns producing a product of 20. Similarly, tiles can be used to model division by splitting the total into equal groups. If dividing 12 by 4, 12 tiles are placed one at a time into 4 groups. The result is 4 groups of 3. This is also an effective method for visualizing the concept of remainders.

Representations of objects can be used to expand on the concrete models of operations. Pictures, dots, and tallies can help model these concepts. Utilizing concrete models and representations creates a foundation upon which to build an abstract understanding of the operations.

Determining Unknown Whole Numbers

In solving multi-step problems, the first step is to line up the available information. Then, try to decide what information the problem is asking to be found. Once this is determined, construct a strip diagram to display the known information along with any information to be calculated. Finally, the missing information can be represented by a **variable** (a letter from the alphabet that represents a number) in a mathematical equation that the student can solve.

For example, Delilah collects stickers, and her friends gave her some stickers to add to her current collection. Joe gave her 45 stickers, and Aimee gave her 2 times the number of stickers that Joe gave Delilah. How many stickers did Delilah have to start with, if after her friends gave her more stickers, she had a total of 187 stickers?

In order to solve this, the given information must first be sorted out. Joe gives Delilah 45 stickers, Aimee gives Delilah 2 times the number Joe gives (2 × 45), and the end total of stickers is 187.

A strip diagram represents these numbers as follows:

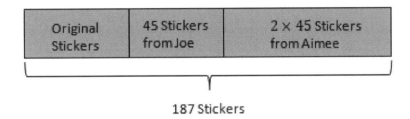

The entire situation can be modeled by this equation, using the variable s to stand for the original number of stickers:

$$s + 45 + (2 \times 45) = 187.$$

Solving for s would give the solution, as follows:

$$s + 45 + 90 = 187$$

$$s + 135 = 187$$

$$s + 135 - 135 = 187 - 135$$

$$s = 52 \text{ stickers.}$$

Word problems take concepts you learned in the classroom and turn them into real-life situations. Some parts of the problem are known and at least one part is unknown. There are three types of instances in which something can be unknown: the starting point, the change, or the final result. These can all be missing from the information they give you.

For solving problems with unknown factors, it is often easiest to set up an array to visualize the grouping of the information. In these problems, set up the initial numbers in uniformly sized groups, so the solution can be determined by inspection of the grouping.

Find the missing number (?) in the following equation:

$$? \times 5 = 35$$

Knowing that one of the factors is to be multiplied is 5 allows the groupings to be made in sets of five columns. In this case, there 5 columns of items are created, until the desired number (35) is reached.

Here, the number of 35 is reached with the seventh row of items. Therefore, the missing factor is 7.

$$5 \times 7 = 35$$

The same problem could be demonstrated with the equation:

$$5 \times ? = 35$$

This would simply require the information to be grouped into five rows, and items added evenly until the desired number (35) is reached.

Again, the solution is:

$$5 \times 7 = 35.$$

This demonstrates the commutative property of multiplication by showing the missing factor could be the number of rows or the number of columns, and yet result in the same solution.

Properties of Multiplication and the Relationship between Multiplication and Division

Applying Properties of Operations as Strategies to Multiply and Divide

Properties of operations exist that make calculations easier. The following table summarizes commonly used properties of real numbers.

Property	Addition	Multiplication
Commutative	$a + b = b + a$	$a \times b = b \times a$
Associative	$(a + b) + c = a + (b + c)$	$(a \times b) \times c = a \times (bc)$
Distributive	$a(b + c) = ab + ac$	

The commutative property of multiplication states that the order in which numbers are multiplied does not change the product. The associative property of multiplication states that the grouping of numbers being multiplied does not change the product. The commutative and associative properties are useful for performing calculations. For example, $(7 \times 5) \times 3$ is equivalent to $(7 \times 3) \times 5$, which is easier to calculate.

The distributive property states that multiplying a sum (or difference) by a number produces the same result as multiplying each value in the sum (or difference) by the number and adding (or subtracting) the products. Consider the following scenario: You are buying three tickets for a baseball game. Each ticket costs $18. You are also charged a fee of $2 per ticket for purchasing the tickets online. The cost is calculated:

$$3 \times 18 + 3 \times 2$$

Using the distributive property, the cost can also be calculated $3(18 + 2)$.

A method for multiplying a two-digit number by a one-digit number is **partial products.** This involves decomposing the larger number into its place values, multiplying the smaller number by these parts, and then adding the totals.

For example, to multiply 64×8, decompose 64 into its place values.

This yields $60 + 4$. Next, multiply 8 by each part:

$$60 \times 8 = 480$$

$$4 \times 8 = 32$$

Now add those totals to find the final answer:

$$480 + 32 = 512$$

This breakdown can also be done as mental math in some situations. For example, to multiply 78×4, rounding 78 up 2 digits to 80 would result in $80 \times 4 = 320$. To finish the problem, the difference from the rounding must be subtracted from the total. Because $80 - 2 = 78$, the difference is 2.

$$2 \times 4 = 8$$

Subtract 8 from the total, $320 - 8 = 312$.

Another quick method to multiply larger numbers involves an *algorithm* to line up the products.

For example, multiply 71×3.

First, line up the numbers on the far right:

$$\begin{array}{r} 71 \\ \times \quad 3 \\ \hline \end{array}$$

Next, beginning on the far right and then moving one place at a time to the left, multiply the two numbers and write the answer below the line in the column of the top number. Continue the process until there are no more numbers to the left on top, as follows:

$$\begin{array}{r} 71 \\ \times \quad 3 \\ \hline 3 \end{array}$$

$$\begin{array}{r} 71 \\ \times \quad 3 \\ \hline 213 \end{array}$$

Some numbers will require a leading number to be carried up to the row to the left, as follows:

$$\begin{array}{r} 7^16 \\ \times \quad 3 \\ \hline 8 \end{array}$$

$$\begin{array}{r} 7^16 \\ \times \quad 3 \\ \hline 228 \end{array}$$

Notice that any carryovers are added to the sum of the two numbers being multiplied.

Understanding Division as an Unknown-Factor Problem

Division is based on dividing a given number into parts. The simplest problem involves dividing a number into equal parts. For example, a pack of 20 pencils is to be divided among 10 children. You would have to divide 20 by 10. In this example, each child would receive 2 pencils.

The symbol for division is $\div$ or /. The equation above is written as $20 \div 10 = 2$, or $20 / 10 = 2$. This means "20 divided by 10 is equal to 2." Division can be explained as the following: for any whole numbers a and b, where b is not equal to zero, $a \div b = c$ if and only if $a = b \times c$. This means, division can be thought of as a multiplication problem with a missing part. For instance, calculating $20 \div 10$ is the same as asking the following: "If there are 20 items in total with 10 in each group, how many are in each group?" Therefore, 20 is equal to ten times what value? This question is the same as asking, "If there are 20 items in total with 2 in each group, how many groups are there?" The answer to each question is 2.

In a division problem, a is known as the **dividend**, b is the **divisor**, and c is the **quotient**. Zero cannot be divided into parts. Therefore, for any nonzero whole number a, $0 \div a = 0$. Also, division by zero is undefined. Dividing an amount into zero parts is not possible.

Multiplication and division are inverse operations. So, multiplying by a number and then dividing by the same number results in the original number. For example, $8 \times 2 \div 2 = 8$ and $12 \div 4 \times 4 = 12$. Inverse operations are used to work backwards to solve problems. If a school's entire 4th grade was divided evenly into 3 classes each with 22 students, the inverse operation of multiplication is used to determine the total students in the grade:

$$22 \times 3 = 66$$

Multiplying and Dividing Within 100

Fluently Multiplying and Dividing Within 100

Mental math is important in multiplication and division. The times tables, for multiplying all numbers from one to 12, should be memorized. This will allow for division within those numbers to be memorized as well. For example, $121 \div 11 = 11$ because it should be memorized that $11 \times 11 = 121$. Here is the multiplication table to be memorized:

x	1	2	3	4	5	6	7	8	9	10	11	12	13	14	15
1	1	2	3	4	5	6	7	8	9	10	11	12	13	14	15
2	2	4	6	8	10	12	14	16	18	20	22	24	26	28	30
3	3	6	9	12	15	18	21	24	27	30	33	36	39	42	45
4	4	8	12	16	20	24	28	32	36	40	44	48	52	56	60
5	5	10	15	20	25	30	35	40	45	50	55	60	65	70	75
6	6	12	18	24	30	36	42	48	54	60	66	72	78	84	90
7	7	14	21	28	35	42	49	56	63	70	77	84	91	98	105
8	8	16	24	32	40	48	56	64	72	80	88	96	104	112	120
9	9	18	27	36	45	54	63	72	81	90	99	108	117	126	135
10	10	20	30	40	50	60	70	80	90	100	110	120	130	140	150
11	11	22	33	44	55	66	77	88	99	110	121	132	143	154	165
12	12	24	36	48	60	72	84	96	108	120	132	144	156	168	180
13	13	26	39	52	65	78	91	104	117	130	143	156	169	182	195
14	14	28	42	56	70	84	98	112	126	140	154	168	182	196	210
15	15	30	45	60	75	90	105	120	135	150	165	180	195	210	225

The values in gray along the diagonal of the table consist of **perfect squares**. A perfect square is a number that represents a product of two equal integers.

Solving Problems Involving the Four Operations, and Identifying and Explaining Patterns in Arithmetic

Solving Two-Step Word Problems Using the Four Operations

Word and story problems should be presented in many different ways. Word problems are presented into real-world situations. These types of problems are situations in which some parts of the problem are known and at least one part is unknown.

There are three types of instances in which something can be unknown: the starting point, the modification, or the final result can all be missing from the provided information.

- For an addition problem, the **modification** is the quantity of the new amount added to the starting point.

- For a subtraction problem, the **modification** is the quantity taken away from the starting point.

Keywords in the word problems can signal what type of operation needs to be used to solve the problem. Words such as *total, increased, combined,* and *more* indicate that addition is needed. Words such as *difference, decreased,* and *minus* indicate that subtraction is needed.

Regarding addition, the given equation is $3 + 7 = 10$.

The number 3 is the starting point. 7 is the modification, and 10 is the result from adding a new amount to the starting point. Different word problems can arise from this same equation, depending on which value is the unknown. For example, here are three problems:

- If a student had three pencils and was given seven more, how many would he have in total?
- If a student had three pencils and a student gave him more so that he had ten in total, how many were given to him?
- A student was given seven pencils so that he had ten in total. How many did he start with?

All three problems involve the same equation, and determining which part of the equation is missing is the key to solving each word problem. The missing answers would be 10, 7, and 3, respectively.

In terms of subtraction, the same three scenarios can occur. The given equation is $6 - 4 = 2$.

The number 6 is the starting point. 4 is the modification, and 2 is the new amount that is the result from taking away an amount from the starting point. Again, different types of word problems can arise from this equation. For example, here are three possible problems:

- *If a student had six quarters and four were taken away, how many would be left over?*

- *If a student had six quarters, purchased a pencil, and had two quarters left over, how many quarters did she pay with?*

- *If a student paid for a pencil with four quarters and had two quarters left over, how many did she have to start with?*

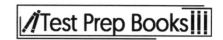

The three question types follow the structure of the addition word problems, and determining whether the starting point, the modification, or the final result is missing is the goal in solving the problem. The missing answers would be 2, 4, and 6, respectively.

The three addition problems and the three subtraction word problems can be solved by using a picture, a number line, or an algebraic equation. If an equation is used, a question mark can be utilized to represent the unknown quantity. For example, $6 - 4 = ?$ can be written to show that the missing value is the result. Using equation form visually indicates what portion of the addition or subtraction problem is the missing value.

Similar instances can be seen in word problems involving multiplication and division. Key words within a multiplication problem involve *times, product, doubled,* and *tripled.* Key words within a division problem involve *split, quotient, divided, shared, groups,* and *half.* Like addition and subtraction, multiplication and division problems also have three different types of missing values.

Multiplication consists of a specific number of groups having the same size, the quantity of items within each group, and the total quantity within all groups. Therefore, each one of these amounts can be the missing value.

For example, the given equation is $5 \times 3 = 15$.

5 and 3 are interchangeable, so either amount can be the number of groups or the quantity of items within each group. 15 is the total number of items. Again, different types of word problems can arise from this equation. For example, here are three problems:

- If a classroom is serving 5 different types of apples for lunch and has three apples of each type, how many total apples are there to give to the students?
- If a classroom has 15 apples of 5 different types, how many of each type are there?
- If a classroom has 15 apples with 3 of each type, how many types are there to choose from?

Each question involves using the same equation to solve, and it is imperative to decide which part of the equation is the missing value. The answers to the problems are 15, 3, and 5, respectively.

Similar to multiplication, division problems involve a total amount, a number of groups having the same size, and a number of items within each group. The difference between multiplication and division is that the starting point is the total amount, which then gets divided into equal quantities.

For example, the equation is $15 \div 5 = 3$.

15 is the total number of items, which is being divided into 5 different groups. In order to do so, 3 items go into each group. Also, 5 and 3 are interchangeable, so the 15 items could be divided into 3 groups of 5 items each. Therefore, different types of word problems can arise from this equation depending on which value is unknown. For example, here are three types of problems:

- A student needs 48 pieces of chalk. If there are 8 pieces in each box, how many boxes should he buy?

- A student has 48 pieces of chalk. If each box has 6 pieces in it, how many boxes did he buy?

- A student has partitioned all of his chalk into 8 piles, with 6 pieces in each pile. How many pieces does he have in total?

Each one of these questions involves the same equation, and the third question can easily utilize the multiplication equation $8 \times 6 = ?$ instead of division. The answers are 6, 8, and 48, respectively.

Calculations relating to real-world expenses rarely have whole number solutions. It is good practice to be able to make one-step calculations that model real-world situations. For example, Amber spends $54 on pet food in a month. If there are no increases in price for the next 11 months, how much will Amber spend on pet food during those 11 months?

Set up the multiplication problem to see how much $54 of food times 11 is:

$$
\begin{array}{r}
54 \\
\times\, 11 \\
\hline
54 \\
+\,540 \\
\hline
594
\end{array}
$$

Amber would spend $594 on pet food over the next 11 months.

As another example, what if Amber had $650; how many months would this last her for pet food expenses if they were $54 per month?

Set the problem up as a division problem to see how many times 54 could divide into 650. The answer will give the number of complete months the expenses could be covered.

$$
\begin{array}{r}
12 \\
54\,\overline{)650} \\
-\,54 \\
\hline
110 \\
-\,108 \\
\hline
2
\end{array}
$$

The remainder does not represent a full month of expenses. Therefore, Amber's $650 would last for a full 12 months of pet food expenses.

Identifying Arithmetic Patterns

Patterns are an important part of mathematics. Identifying and understanding how a group or pattern is represented in a problem is essential for being able to expand this process to more complex problems. A simple input-output table can model a pattern that pertains to a specific situation or equation. These can then be utilized in other areas in math, such as graphing.

For example, for every 1 parakeet the pet store sells, it sells 5 goldfish. Using the following equation to model this situation, fill in numbers missing in the input-output table, to show the total number of pets sold by the store.

Total number of pets sold (t) = number of parakeets (p) + number of parakeets (p) × 5 goldfish

$$t = p + (p \times 5)$$
$$t = 6p$$

p	t
1	6
2	12
3	
4	24
5	

The missing numbers are 18 and 30.

This can also be shown by using an equation. If 3 is put in for p, it would look as follows:

$$t = 6 \times 3$$

$$t = 18$$

If 5 is put in for p, it would look as follows:

$$t = 6 \times 5$$

$$t = 30$$

The completed table would appear as follows:

p	t
1	6
2	12
3	18
4	24
5	30

By looking at the completed table, the numeric patterns between consecutive p and t values (from one row to the next) can be seen. The p-values increase by 1, and the t-values increase by 6.

Number and Operations in Base Ten

Using Place Value Understanding and Properties of Operations to Perform Multi-Digit Arithmetic

Rounding Whole Numbers

When numbers are counted, it is really counting groups of 10. That number is consistent throughout the set of natural numbers, whole numbers, etc., and is referred to as working within a base 10 numeration system. Only the numbers from 0 to 9 are utilized to represent any number, and the foundation for doing so involves **place value**. Numbers are written side by side, to show the amount is in each place value.

The idea of place value can be better understood by considering how the number 10 is different from 0 to 9. It has two digits instead of just one. The 1 in the first digit is in the tens' place, and the 0 in the second digit is in the ones place. Therefore, there is 1 group of tens and 0 ones. 11 is the next number that can be introduced because this number has 1 ten and 1 one. Considering numbers from 11 to 19 should be the next step. Anytime a new number is introduced, writing out the numbers as *eleven*, *twelve*, etc., should be done simultaneously. Each value within this range of numbers consists of one group of ten and a specific number of leftover ones. Counting by tens can be introduced once the tens column is understood. This process consists of increasing the number in the tens place by one. For example, counting by ten starting at 17 would result in the next four values being 27, 37, 47, and 57.

Later, you can think about higher place values. Base ten blocks can be utilized to help understand place value and assist with counting large numbers. They consist of cubes that help with the visualization of counting and operations.

For example, here is a diagram of base ten blocks representing ones, tens, and a hundred:

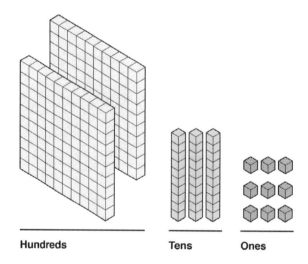

Hundreds   Tens   Ones

Also, a place value chart should be utilized for numbers containing higher digits.

Here is an example of a place value chart:

	MILLIONS			THOUSANDS			ONES			.	DECIMALS		
billions	hundred millions	ten millions	millions	hundred thousands	ten thousands	thousands	hundreds	tens	ones		tenths	hundredths	thousandths

Rounding numbers changes the given number to a simpler and less accurate number than the exact given number. Rounding allows for easier calculations which estimate the results of using the exact given number. The accuracy of the estimate and ease of use depends on the place value to which the number is rounded. Rounding numbers consists of:

- Determining what place value the number is being rounded to
- Examining the digit to the right of the desired place value to decide whether to round up or keep the digit
- Replacing all digits to the right of the desired place value with zeros

To round 746,311 to the nearest ten thousandth, the digit in the ten thousandth place should be located first. In this case, this digit is 4 (7<u>4</u>6,311). Then, the digit to its right is examined. If this digit is 5 or greater, the number will be rounded up by increasing the digit in the desired place by one. If the digit to the right of the place value being rounded is 4 or less, the number will be kept the same. For the given example, the digit being examined is a 6, which means that the number will be rounded up by increasing the digit to the left by one. Therefore, the digit 4 is changed to a 5. Finally, to write the rounded number, any digits to the left of the place value being rounded remain the same and any to its right are replaced with zeros. For the given example, rounding 746,311 to the nearest ten thousand will produce 750,000. To round 746,311 to the nearest hundred, the digit to the right of the three in the hundreds place is examined to determine whether to round up or keep the same number. In this case, that digit is a one, so the number will be kept the same and any digits to its right will be replaced with zeros. The resulting rounded number is 746,300.

Rounding to the Nearest 10, 100, or 1000

Estimation is an important tool in mathematics and in science. Understanding how to round numbers and compare and combine them is necessary to be able to properly estimate. For example, Alex's server holds 125,678 movies, while Mim's server holds 332,102 movies. If Alex and Mim have an overlap of 52,032 movies, approximately how many unique movies would the two have if they combined collections?

First, compare the numbers and decide which place value would need to be rounded in order to make an estimate for this problem. Because the smallest number is 52,032, rounding to the ten-thousands place is necessary for proper estimation. Traditional rounding uses one place smaller than the one being rounded to, to determine if the actual place value will be rounded up or remain the same. Therefore, if rounding to the nearest ten, look at the digit in the ones place. To round to the nearest hundred, look at the digit in the tens place, and to round to the nearest thousand, look at the digit in the hundreds place. If the number in the designated place is 5 or greater, the digit in question will increase by one. If the number is less than 5, the digit in question will remain the same.

In the case of 52,032:

Look at the thousands place:

52,032   Since 2 is less than 5, the 5 in the ten-thousands place will remain the same.

All of the numbers after the ten-thousands place now become zeroes:

50,000

To round the other two numbers to the appropriate place, the rounding will look like the following:

125,678   Since 5 is greater than or equal to 5, the value in the ten-thousands place increases by 1.

130,000

332,102   Since 2 is less than 5, the value in the ten-thousands place remains the same.

330,000

An estimate would take the following calculation:

$330,000 + 130,000 - 50,000 = 410,000$

There are approximately 410,000 unique movies in the combined collection.

When rounding up, if the digit to be increased is a 9, the digit to its left is increased by 1 and the digit in the desired place value is changed to a zero. For example, the number 1,598 rounded to the nearest ten is 1,600. Another example shows the number 467,296 rounded to the nearest hundred thousand is 500,000 because the 6 in the ten-thousands place indicates that we need to round up. This changes the 4 in the hundred-thousandth place to a 5.MAFS.3.NBT.1.2 Fluently add and subtract within 1,000 using strategies and algorithms based on place value, properties of operations, and/or the relationship between addition and subtraction.

Multiplying One-Digit Whole Numbers by Multiples of 10

When a whole one-digit whole number (1, 3, 4, 8, etc.) is multiplied by a multiple of 10, such as 10, 20, 30, or 50, the one-digit whole number factor is multiplied by the first digit of the multiple of 10 (such as 1 in 10, 2 in 20, 3 in 30), and then a zero is added to end of the product. For example, when considering 4×10, the 4 is multiplied by 1 to get 4, and then one zero is added to the end, which results in 40. Consider multiplying 8×60. A quick way to find the product using mental math is again to multiply 8×6. Using memorized times table facts, the result of 48 is quickly obtained. Then, the zero is added to the end to get 480.

Number and Operations — Fractions

Fractions as Numbers

Explaining Fractions as One Part of a Whole and Composing and Decomposing Them

A **fraction** is a part of something that is whole. Items such as apples can be cut into parts to help visualize fractions. If an apple is cut into 2 equal parts, each part represents ½ of the apple. If each half is cut into two parts, the apple now is cut into quarters. Each piece now represents ¼ of the apple. In this example, each part is equal because they all have the same size.

For example, a circle can be drawn and divided it into 6 equal parts:

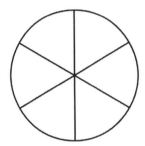

Shading can be used to represent parts of the circle that can be translated into fractions. The top of the fraction, the **numerator,** can represent how many segments are shaded. The bottom of the fraction, the **denominator,** can represent the number of segments that the circle is broken into. A pie is a good analogy to use in this example. If one piece of the circle is shaded, or one piece of pie is cut out, $1/_6$ of the object is being referred to. An apple, a pie, or a circle can be utilized in order to compare simple fractions. For example, showing that $1/_2$ is larger than $1/_4$ and that $1/_4$ is smaller than $1/_3$ can be accomplished through shading. A **unit fraction** is a fraction in which the numerator is 1, and the denominator is a positive whole number. It represents one part of a whole—one piece of pie.

Fractions can be broken apart into sums of fractions with the same denominator. For example, the fraction $\frac{5}{6}$ can be decomposed into sums of fractions with all denominators equal to 6 and the numerators adding to 5. The fraction $\frac{5}{6}$ is decomposed as:

$$\frac{3}{6} + \frac{2}{6};\ or\ \frac{2}{6} + \frac{2}{6} + \frac{1}{6};\ or\ \frac{3}{6} + \frac{1}{6} + \frac{1}{6};\ or\ \frac{1}{6} + \frac{1}{6} + \frac{1}{6} + \frac{2}{6};\ or\ \frac{1}{6} + \frac{1}{6} + \frac{1}{6} + \frac{1}{6} + \frac{1}{6}$$

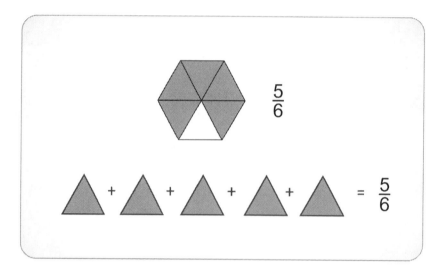

As mentioned, a unit fraction is a fraction in which the numerator is 1. If decomposing a fraction into unit fractions, the sum will consist of a unit fraction added the number of times equal to the numerator. For example, $\frac{3}{4} = \frac{1}{4} + \frac{1}{4} + \frac{1}{4}$ (unit fractions $\frac{1}{4}$ added 3 times). Composing fractions is simply the opposite of decomposing. It is the process of adding fractions with the same denominators to produce a single fraction. For example,

$$\frac{3}{7} + \frac{2}{7} = \frac{5}{7}\ and\ \frac{1}{5} + \frac{1}{5} + \frac{1}{5} = \frac{3}{5}$$

Representing Fractions and Decimals on a Number Line

To represent fractions and decimals as distances beginning at zero on a number line, it's helpful to relate the fraction to a real-world application. For example, a charity walk covers $\frac{3}{10}$ of a mile. How could this distance be represented on a number line?

First, divide the number line into tenths, as follows:

0   0.1   0.2   0.3   0.4   0.5   0.6   0.7   0.8   0.9   1

If each division on the number line represents one-tenth of one, or $\frac{1}{10}$, then representing the distance of the charity walk, $\frac{3}{10}$, would cover 3 of those divisions and look as follows:

So, the fraction $\frac{3}{10}$ is represented by covering from 0 to 0.3 (or 3 sections) on the number line.

Representing a Unit Fraction on a Number Line

A **unit fraction** is a fraction in which the numerator is 1, and the denominator is a positive whole number. It represents one part of a whole—one piece of pie. A unit fraction is sometimes symbolized mathematically by 1/b, where b can be any positive or negative number (called an **integer**). For now, it makes sense to just consider positive whole numbers like 2, 3, 4, or 10. Examples of unit fractions are $\frac{1}{2}$ or $\frac{1}{8}$.

To precisely understand a number being represented on a number line, the first step is to identify how the number line is divided up. When utilizing a number line to represent fractions, it is helpful to label the divisions, or insert additional divisions, as needed.

For example, what fraction is marked by the point on the following number line?

First, figure out how the number line is divided up. In this case, it has 4 sections, so it is divided into fourths. To use this number line with the divisions, label the divisions as follows:

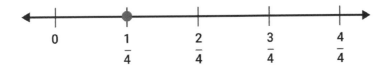

Because the dot is placed on the first section marker, it represents 1 portion out of 4 or $\frac{1}{4}$.

Representing a Fraction *a/b* on a Number Line

A fraction where the numerator (the top number) is not 1 can be a multiple of a unit fraction. For example, $\frac{3}{4}$ is three times the unit fraction $\frac{1}{4}$. These fractions are sometimes expressed mathematically as *a/b.* They can also be expressed on a number line.

Let's look at an example:

What fraction is marked by the point on the following number line?

First, determine what the number line is divided up into and mark it on the line.

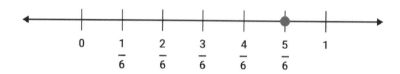

This number line is divided up into six portions so each portion represents $\frac{1}{6}$. The division and labeling of the number line assists in easily reading the dot as marking 5 out of those 6 portions or $\frac{5}{6}$.

Understanding Equivalent Fractions Using a Number Line

Like fractions, or **equivalent fractions**, represent two fractions that are made up of different numbers, but represent the same quantity. For example, the given fractions are $^4/_8$ and $^3/_6$. If a pie was cut into 8 pieces and 4 pieces were removed, half of the pie would remain. Also, if a pie was split into 6 pieces and 3 pieces were eaten, half of the pie would also remain. Therefore, both of the fractions represent half of a pie. When working with fractions in mathematical expressions, like fractions should be simplified. Both $^4/_8$ and $^3/_6$ can be simplified into $^1/_2$.

Equivalent fractions can be visually represented by shading in the fractions on shapes of equivalent areas. For example, if a pie were separated into slices and half of the slices were shaded, the visual would represent equivalent fractions.

$$\frac{3}{6} = \frac{1}{2}$$ $$\frac{2}{4} = \frac{1}{2}$$

$$\frac{4}{8} = \frac{1}{2}$$ $$\frac{5}{10} = \frac{1}{2}$$

On a number line, the representation is visually comparable.

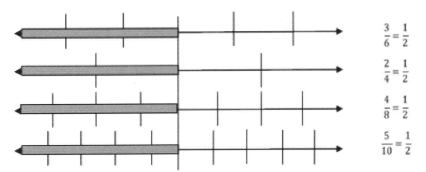

$$\frac{3}{6} = \frac{1}{2}$$

$$\frac{2}{4} = \frac{1}{2}$$

$$\frac{4}{8} = \frac{1}{2}$$

$$\frac{5}{10} = \frac{1}{2}$$

Both sets of diagrams represent the same fractions. Equivalence is true only if the exact same amount in each area is shaded.

Identifying Equivalent Fractions

Like fractions, or **equivalent fractions**, represent two fractions that are made up of different numbers, but represent the same quantity. For example, the given fractions are $^4/_8$ and $^3/_6$. If a pie was cut into 8 pieces and 4 pieces were removed, half of the pie would remain. Also, if a pie was split into 6 pieces and 3 pieces were eaten, half of the pie would also remain. Therefore, both of the fractions represent half of a pie. These two fractions are referred to as like fractions. **Unlike fractions** are fractions that are different and cannot be thought of as representing equal quantities. When working with fractions in mathematical expressions, like fractions should be simplified. Both $^4/_8$ and $^3/_6$ can be simplified into $^1/_2$.

Comparing fractions can be completed through the use of a number line. For example, if $\frac{3}{5}$ and $\frac{6}{10}$ need to be compared, each fraction should be plotted on a number line. To plot $\frac{3}{5}$, the area from 0 to 1 should be broken into 5 equal segments, and the fraction represents 3 of them. To plot $\frac{4}{10}$, the area from 0 to 1 should be broken into 10 equal segments, and the fraction represents 6 of them.

It can be seen that $\frac{3}{5} = \frac{6}{10}$

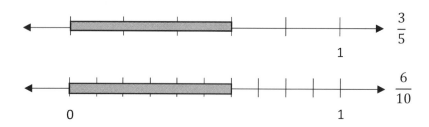

Like fractions are plotted at the same point on a number line.

Recognizing and Representing Equivalent Fractions Using Objects and Models

Grouping objects or shading equivalent shapes are helpful in representing equivalent fractions. For example, if the fraction $\frac{1}{2}$ were to be represented by grouping objects totaling 6 and objects totaling 8, it can be shown through the following groupings:

$\frac{1}{2}$ of 6 is 3, or $\frac{3}{6}$:

$\frac{1}{2}$ of 8 is 4, or $\frac{4}{8}$:

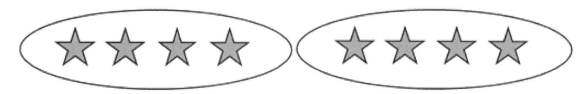

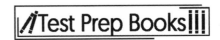
Area models can also be used to represent equivalent fractions. For example, each of the following area models contain 36 small squares with 6 shaded in. This is a fraction of $\frac{6}{36}$ or $\frac{1}{6}$ of the total number of squares, even though the arrangement of shaded squares is different.

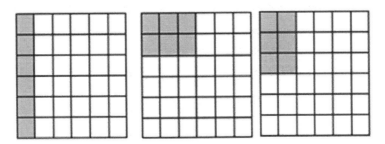

Expressing Whole Numbers as Fractions, and Recognize Fractions That are Equivalent to Whole Numbers

A fraction can be thought of as the quotient to a division problem. The numerator is divided by the top number. For example, $\frac{1}{3}$ is the quotient of $1 \div 3$. Therefore, recalling that any number divided by 1 is itself (for example, $7 \div 1 = 7$), whole numbers can be written as fractions that have the whole number as the numerator and 1 as the denominator. This is because this would be the same as dividing that whole number by 1. As an example, the whole number 7 is written in fractional form as $\frac{7}{1}$.

Comparing Two Fractions

To compare fractions with either the same **numerator** (top number) or same **denominator** (bottom number), it is easiest to visualize the fractions with a model.

For example, which is larger, $\frac{1}{3}$ or $\frac{1}{4}$? Both numbers have the same numerator, but a different denominator. In order to demonstrate the difference, shade the amounts on a pie chart split into the number of pieces represented by the denominator.

The first pie chart represents $\frac{1}{3}$, a larger shaded portion, and is therefore a larger fraction than the second pie chart representing $\frac{1}{4}$.

If two fractions have the same denominator (or are split into the same number of pieces), the fraction with the larger numerator is the larger fraction, as seen below in the comparison of $\frac{1}{3}$ and $\frac{2}{3}$:

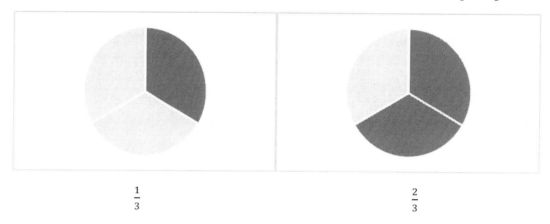

$$\frac{1}{3}$$

$$\frac{2}{3}$$

A **unit fraction** is one in which the numerator is 1 ($\frac{1}{2}, \frac{1}{3}, \frac{1}{8}, \frac{1}{20}$, etc.). The denominator indicates the number of equal pieces that the whole is divided into. The greater the number of pieces, the smaller each piece will be. Therefore, the greater the denominator of a unit fraction, the smaller it is in value. Unit fractions can also be compared by converting them to decimals. For example, $\frac{1}{2} = 0.5$, $\frac{1}{3} = 0.\overline{3}$, $\frac{1}{8} = 0.125$, $\frac{1}{20} = 0.05$, etc.

Comparing two fractions with different denominators can be difficult if attempting to guess at how much each represents. Using a number line, blocks, or just finding a common denominator with which to compare the two fractions makes this task easier.

For example, compare the fractions $\frac{3}{4}$ and $\frac{5}{8}$.

The number line method of comparison involves splitting one number line evenly into 4 sections, and the second number line evenly into 8 sections total, as follows:

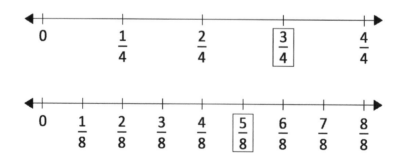

Here it can be observed that $\frac{3}{4}$ is greater than $\frac{5}{8}$, so the comparison is written as $\frac{3}{4} > \frac{5}{8}$.

This could also be shown by finding a common denominator for both fractions, so that they could be compared. First, list out factors of 4: 4, 8, 12, 16.

Then, list out factors of 8: 8, 16, 24.

Both share a common factor of 8, so they can be written in terms of 8 portions. In order for $\frac{3}{4}$ to be written in terms of 8, both the numerator and denominator must be multiplied by 2, thus forming the new fraction $\frac{6}{8}$. Now the two fractions can be compared.

Because both have the same denominator, the numerator will show the comparison.

$$\frac{6}{8} > \frac{5}{8}$$

Remember to use the inequality symbols when representing unequal comparisons:

Symbol	Phrase
<	is under, is below, smaller than, beneath
>	is above, is over, bigger than, exceeds
≤	no more than, at most, maximum
≥	no less than, at least, minimum

Measurement and Data

Solving Problems Involving Measurement and Estimation of Intervals of Time, Liquid Volumes, and Masses of Objects

Telling Time and Solving Problems Related to Time

Telling Time

On a clock, there is a long hand and a short hand. The longer and larger hand is known as the **minute hand**, and it points to the minute associated with the current time. The minute hand goes around the clock in a clockwise direction every sixty minutes, or one hour. Remember, one hour is sixty minutes long. There are sixty minutes on a clock, but they are not numbered individually. However, there are tick marks along the circumference of the circle (the outline). There is one tick mark for each minute. The smaller and shorter hand is known as the **hour hand**, and it points to the hour associated with the current time. The hour hand goes around the clock in a clockwise direction every twelve hours, or two times per day, because a full day (and night) is 24 hours. There are twelve hours on the clock, and the numbers are visible. If the hour hand is just past one of the hours, it is associated with that hour. For instance, consider the following example:

The minute hand is pointing to the 45th minute and the hour hand is past 9 but before 10, so the time is 9:45. It could be either a.m. (morning, before noon) or p.m. (afternoon or evening).

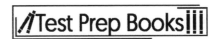

Because there are 60 tick marks on the clock (one per minute) and 12 labeled numbers for the hours, every number is separated by five tick marks (60 ÷ 12 = 5). Therefore, you can memorize the equivalent minutes for each number so that when the large minute hand is on a number, you automatically know how many minutes past the hour it is. Here is a list:

- 1 = 5 minutes
- 2 = 10 minutes
- 3 = 15 minutes
- 4 = 20 minutes
- 5 = 25 minutes
- 6 = 30 minutes
- 7 = 35 minutes
- 8 = 40 minutes
- 9 = 45 minutes
- 10 = 50 minutes
- 11 = 55 minutes
- 12 = 0 minutes

Solving Problems Related to Time

Time is measured in units such as *seconds, minutes, hours, days,* and *years*. For example, there are 60 seconds in a minute, 60 minutes in each hour, and 24 hours in a day.

When dealing with problems involving elapsed time, break the problem down into workable parts. For example, suppose the length of time between 1:15 p.m. and 3:45 p.m. must be determined. From 1:15 p.m. to 2:00 p.m. is 45 minutes (knowing there are 60 minutes in an hour). From 2:00 p.m. to 3:00 p.m. is 1 hour. From 3:00 p.m. to 3:45 p.m. is 45 minutes. The total elapsed time is 45 minutes plus 1 hour plus 45 minutes. This sum produces 1 hour and 90 minutes. 90 minutes is over an hour, so this is converted to 1 hour (60 minutes) and 30 minutes. The total elapsed time can now be expressed as 2 hours and 30 minutes.

To illustrate time intervals, a clock face can show solutions.

For example, Ani needs to complete all of her chores by 1:50 p.m. If she begins her chores at 1:00 p.m., can she finish the following? Vacuuming (15 minutes), dusting (10 minutes), replacing light bulbs (5 minutes), and degreasing the garage floor (25 minutes).

A blank clock face is useful in illustrating the time lapse necessary for all of Ani's tasks.

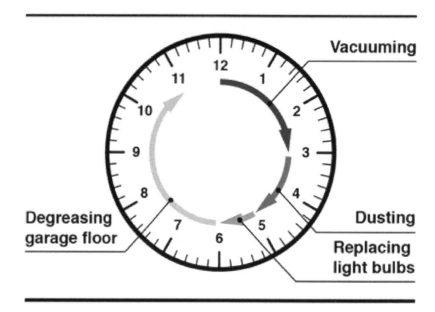

It is easy to see that the chores will span beyond 50 minutes after the hour, so no, Ani could not complete the chores in the given time frame.

Determining Appropriate Measurements of Liquid Volume and Weight

Measurement is how an object's length, width, height, weight, and so on, are quantified. Measurement is related to counting, but it is a more refined process.

Weight units can vary, based on whether the substance being measured is a liquid or a solid. Standard units of weight to measure liquids include *ounces*, *pints*, *quarts*, and *gallons*. Occasionally, solids can also be measured using pints and quarts. For example, both milk and berries can be measured in pints. Other units of weight are *pounds* and *tons*.

The proper instruments for measurements depend upon the units being measured. The following instruments are used for measuring the listed values, along with the specific units they measure:

- Volume: Measuring cup (fluid ounces, cups), graduated cylinder (cubic centimeters, milliliters), beaker (milliliters)

- Weight: Scale (pounds, Newtons)

For example, with what instrument would Mart measure a cup of sugar he wants to combine with other ingredients to make a pie?

The correct answer is a measuring cup.

If Mart needed to measure large quantities of liquid, he could use a quart, pint, or gallon measuring container. The conversions between cups and quarts is four cups per one quart, and between quarts and gallons, it's four quarts per one gallon.

If Mart needed 2 quarts of liquid for a recipe and only has a measuring cup, how could he measure out 2 quarts?

The solution would involve Mart measuring out 2 quarts by filling the cup 8 times.

Representing and Interpreting Data

Drawing a Scaled Picture Graph and a Scaled Bar Graph to Represent a Data Set

Graphs can be constructed to represent data.

A **picture graph** is a diagram that shows pictorial representation of data being discussed. The symbols used can represent a certain number of objects. Notice how each fruit symbol in the following graph represents a count of two fruits. One drawback of picture graphs is that they can be less accurate if each symbol represents a large number. For example, if each banana symbol represented ten bananas, and students consumed 22 bananas, it may be challenging to draw and interpret two and one-fifth bananas as a frequency count of 22.

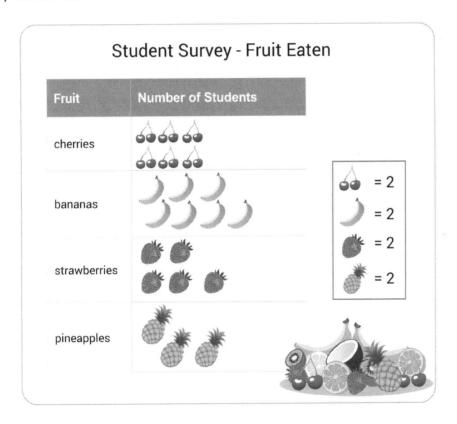

To draw a picture graph to scale, the pictures or symbols chosen to represent the items in the various categories must represent an equal number of items. For example, in the graph above, each picture of one of the fruit types represents two of that fruit. The **key** is the box on the side that shows readers what the symbols represent. This allows the data to be counted.

A **bar graph** displays the number of data points (on the up-and-down (vertical) axis, called the *y*-axis) for the items listed along the bottom on the *x*-axis. These items are considered categorical data. Therefore, the horizontal axis represents each category and the vertical axis represents the frequency for the category. A bar is drawn for each category (often different colors) with a height extending to the frequency for that category within the data set. A double bar graph displays two sets of data that contain data points consisting of the same categories. The double bar graph below indicates that two girls and four boys like Pad Thai the most out of all the foods, two boys and five girls like pizza, and so on.

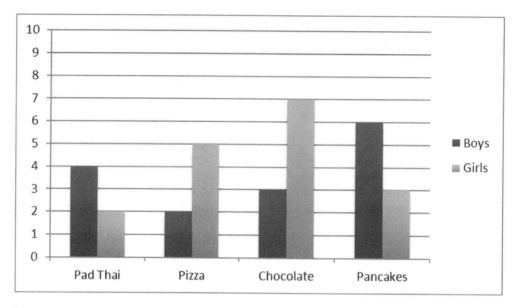

When drawing a bar graph, it is important to draw it to scale. This involves making sure the spacing between the numbers on the *y*-axis (the vertical one) are equal and that the interval between numbers is the same. A ruler or graph paper can be used to make this process neater and easier. If each box on the graph paper represents one, you must keep that same relationship all the way up the graph (a box cannot suddenly jump 2, for example).

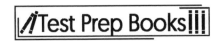

Solving Problems Using Data

Multi-step problems can be solved using information displayed in a stem-and-leaf plot. For example, the following graph shows the data collected regarding snowfall on top of specific mountains in the Alps.

It can be used to answer the following questions regarding the data.

February		April
	0	2, 4, 9
8	1	
9, 5, 2	2	2, 4, 7, 8
6, 2	3	1, 3
9, 6, 2	4	0, 3, 6, 8, 9
8, 7, 6, 6, 5	5	0
4	6	2

On the left side, 6 | 4 means 4.6 in.   On the right side, 4 | 6 means 4.6 in.

1. Which month had the largest collective snowfall, February or April?
2. How much larger was this snowfall?
3. How many mountains reported more than 4.0 inches of snowfall in February?
4. What is the difference between the lowest reported snowfall in February and the lowest reported snowfall in April?
5. What was the total for the three highest snowfalls in April?

The solution involves adding up the total amount of snowfall in both months individually and finding that February reported more snowfall than April with a total of 64.5 inches. This total was more than the April snowfall by 12.7 inches. There are 9 data points that are higher than 4.0 inches in February. The lowest reported snowfall in February is 1.8 inches, and the lowest reported snowfall in April is 0.2 inches. The difference between the two points is 1.6 inches. The three highest snowfalls in April are 6.2, 5.0, and 4.9. The total of these is 16.1 inches.

Let's try a question that uses a bar graph. This graph was created after a science experiment, where third grade students planted seeds in different pots. They made three experimental groups that received a different amount of fertilizer and then the height of the plants was measured after one month.

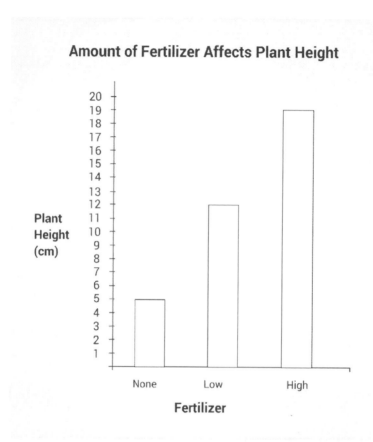

Looking at the graph, how much taller did the plants in the "high" fertilizer group grow than the ones in the "low" fertilizer group?

To answer this question, the height of the high group is checked by tracing a finger or lining up a piece of paper or ruler over to the numbers on the y-axis. Then, it can be seen that these plants grew to be 19 cm. The same procedure is done for the "low" group; they grew 12 cm. Then, a subtraction problem is set up to find the difference: 19 – 12 = 7cm. Therefore, the plants in the "high" fertilizer group grow 7cm more than the ones in the "low" fertilizer group.

Generating Measurement Data by Measuring Lengths Using Rulers

We can measure items in different ways. For instance, we can measure the length, mass, or density of something. In each case, we used a different tool. For length, we can use rulers and the inch is the standard unit of measurement on a ruler. The ruler is labeled with numbers, and in the United States, those labels usually represent inches. (Some rulers have centimeters on the other side as well.) If you need to measure something to the nearest inch, you use the numbers shown on the ruler. Also, you can use the smaller lines in between each inch if you need to measure objects to the nearest half inch or quarter inch. Two half inches make a whole inch. Therefore, if you are measuring to the nearest half

inch, you use the halfway mark. Items typically don't stop exactly at a single mark on the ruler, so you use the mark that is closest to the edge of the item.

If we want to measure to the nearest quarter inch, think of an inch separated into four equal parts. Each equal part is a quarter inch. The object can be measured as one-quarter, two-quarters (one-half inch), three-quarters, or four-quarters of an inch (one inch). Here is an example of a line plot showing one inch, where each half and quarter inch partitioned:

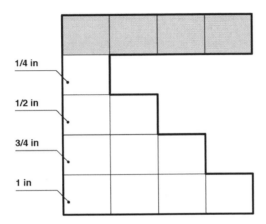

A line plot can be used to display a data set of measurements in fractions of units as well. The reason why this type of plot is necessary would be to organize data of different measurements in a real-world situation. Not every measurement will be a whole number. For instance, let's say you collected 15 leaves from the backyard and measured them to the nearest quarter inch. Your measurements were 3 quarter-inch leaves, 6 half-inch leaves, 5 three quarter-inch leaves, and 1 one-inch leaves. The line plot that represents this data is shown here:

Leaf Lengths

	✗		
	✗	✗	
	✗	✗	
✗	✗	✗	
✗	✗	✗	
✗	✗	✗	✗
$\frac{1}{4}$ in	$\frac{1}{2}$ in	$\frac{3}{4}$ in	1 in

Note that the number line shows the four different measurement possibilities and the x's represent the frequency, or number of times, that each measurement appeared. This line plot can be used to answer questions about the data set. For instance, the tallest height at the half-inch mark shows that this was the most frequent found measurement. Also, the chart could be used to find the difference in lengths between the longest and shortest leave found, which is called the **range** of the measurements. The longest leaf was 1 inch and the shortest leaf was a quarter-inch, so the difference is found through subtraction, with a result of $\frac{3}{4}$ inch.

Geometric Measurement: Understanding Concepts of Area and Relating Area to Multiplication and Addition

Area as an Attribute of Plane Figures

The Unit Square

A **unit square** is a tool that can be used to measure area. If you are measuring a small area, your unit square could be 1 centimeter by 1 centimeter or 1 inch by 1 inch. This means each side is 1 unit in length. For larger areas, your unit square could be 1 foot by 1 foot, 1 yard by 1 yard, or 1 meter by 1 meter.

For instance, consider the situation where you need to tile your kitchen floor. You must purchase square tiles that are 1 foot long on each side. You would need to know how many tiles you need, so to do this, you would partition the floor into unit squares. Let's say your floor has the following shape, which has already been segmented into 1 ft by 1 ft squares:

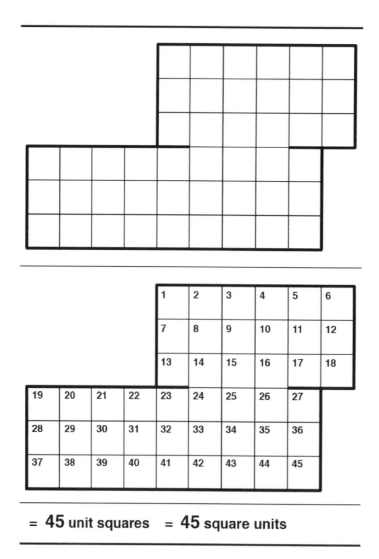

= **45** unit squares   = **45** square units

There are 45 squares, which means that you would need 45 tiles to retile the floor. We have used the process of counting unit squares for the real-life situation of floor tiling.

Using the Unit Square to Determine Area of a Plane Figure

In the previous example, we split up the kitchen floor, which had no gaps or overlaps, into 45 unit squares that had lengths of 1 foot on each side. Because we had 45 unit squares, we can say that the floor had an area of 45 square feet. We are counting the number of squares, so the units change from feet, which is one-dimensional, to square feet, which is two-dimensional.

The scenario of determining area with unit squares does not have to be completely square or include only full squares. For example, consider the following shape:

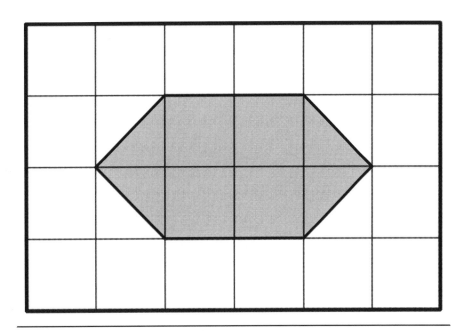

4 full squares + 4 half square = 6 unit squares

There are 4 unit squares in the center, and on the sides, there are four segments that are exactly half of a unit square. If you add those half-segments together, the result is two whole unit squares. Therefore, the shape has 6 total unit squares (4 full squares plus 2 full squares formed by the 4 half-squares), so it has an area of 6 square units. Because the picture does not specify the type of units (such as feet, inches, centimeters, etc.), we just use "square units" to represent the units. Remember, it is always important to indicate what units are used when recording and reporting measurements. If no specific unit is mentioned, like in this case, you write "square units" (if it's area, like in this example), "units" (if it's a one-dimensional measurement, like length), or "cubic units" (if it's volume, like in a cube or shoe box).

Measuring Areas

The area of a two-dimensional figure refers to the number of square units needed to cover the interior region of the figure. This concept is similar to wallpaper covering the flat surface of a wall. For example, if a rectangle has an area of 10 square centimeters (written $10cm^2$), it will take 10 squares, each with sides one centimeter in length, to cover the interior region of the rectangle. Note that area is measured in square units such as: square centimeters or cm^2; square feet or ft^2; square yards or yd^2; square miles or mi^2.

Relating Area to the Operations of Multiplication and Addition

Finding the Area of a Rectangle with Whole-Number Side Lengths by Tiling It

Using the unit square, the area of a rectangle can be calculated by tiling it with unit squares and then counting the squares. As an example, let's consider a rectangle with a length of 4 units and a width of 3 units. The goal is to know how many square units there are in total (the area of the rectangle). Three rows of four squares gives $4 + 4 + 4 = 12$. Also, three times four squares gives $3 \times 4 = 12$. Therefore, for any whole numbers a and b, where a is not equal to zero, $a \times b = b + b + \cdots b$, where b is added a times. Also, $a \times b$ can be thought of as the number of units in a rectangular block consisting of a rows and b columns. For example, 3×7 is equal to the number of squares in the following rectangle:

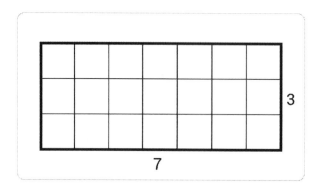

The answer is 21, and there are 21 squares in the rectangle. The area of the rectangle, therefore, is also 21 square units.

Multiplying Side Lengths to Find Areas of Rectangles

The area of a rectangle is found by multiplying its length, l, times its width, w. Therefore, the formula for area is $A = l \times w$. An equivalent expression is found by using the term base, b, instead of length, to represent the horizontal side of the shape. In this case, the formula is $A = b \times h$. This same formula can be used for all parallelograms.

Here is a visualization of a rectangle with its labeled sides:

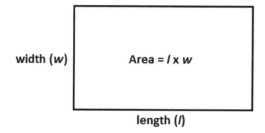

Using Models for Area

The concept of area can be related to the operations of multiplication and addition. Multiplication can be thought of as repeated addition. To model the relationship between area and these operations, organizing objects into arrays or groups of equal numbers for combining is helpful.

For example, a jewelry store's sales are represented by ring boxes in the following diagram. How many total sales were there?

The array above of three rows by four columns (or 3 x 4) shows a total of 12 boxes. Notice that the array can be seen as comprising either three rows of four boxes each, or four columns of three boxes each. This array is akin to a model to calculate the area of a rectangle that is 3 x 4.

Here is another way to look at the jewelry boxes, but with four rows and three columns (a 4 x 3 array).

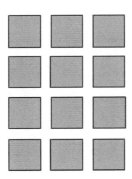

These two diagrams demonstrate the **commutative property of multiplication**, the idea that numbers can be multiplied in any order.

Another method of multiplication can be done with the use of an **area model**. An area model is a rectangle that is divided into rows and columns that match up to the number of place values within each number. Take the example 29×65. These two numbers can be split into simpler numbers: $29 = 25 + 4$ and $65 = 60 + 5$. The products of those 4 numbers are found within the rectangle and then summed up to get the answer. The entire process is:

$$(60 \times 25) + (5 \times 25) + (60 \times 4) + (5 \times 4) = 1,500 + 240 + 125 + 20 = 1,885$$

Here is the actual area model:

	25	**4**
60	60x25 1,500	60x4 240
5	5x25 125	5x4 20

```
    1 , 5 0 0
        2 4 0
        1 2 5
+        2 0
    1 , 8 8 5
```

Multiplication facts can also be represented using rectangular arrays. **Rectangular arrays** include an arrangement of rows and columns that correspond to the factors and display product totals. Again, each one of these dots can be thought of as a square unit; thus, the rectangular array helps calculate the area of the plot of dots.

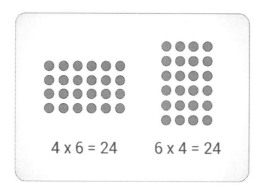

4 x 6 = 24   6 x 4 = 24

Recognizing Area as Additive and Finding Areas of Rectilinear Figures

To understand that oddly formed shapes can be decomposed into more familiar shapes, it is important to visualize their decomposition into the more familiar shapes. Showing the area of the decomposed shapes to see the total area of an odd shape can help with understanding this comparison.

For example:

To plan a garden, Mark needs to separate out the entire garden in three rectangles for planting. Then, he must calculate the area of the entire shape below.

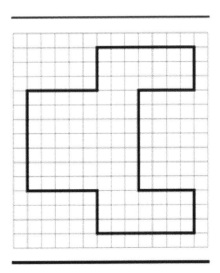

First, divide the figure into three rectangles as follows:

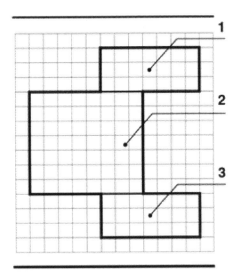

Next, calculate the area of each rectangle by counting how many squares it contains.

Rectangle 1 = 8 squares

Rectangle 2 = 16 squares

Rectangle 3 = 8 squares

Totaling all three rectangles results in the following:

$$8 + 16 + 8 = 32 \ square \ units$$

The total area of the garden plot is the same as the sum of the areas of the decomposed rectangles.

Geometric Measurement: Perimeter is an Attribute of Plane Figures

Solving Problems Related to Perimeter

Perimeter is the length of all its sides. The perimeter of a given closed sided figure would be found by first measuring the length of each side and then calculating the sum of all sides. A rectangle consists of two sides called the length (l), which have equal measures, and two sides called the width (w), which have equal measures. Therefore, the perimeter (P) of a rectangle can be expressed as:

$$P = l + l + w + w$$

This can be simplified to produce the following formula to find the perimeter of a rectangle:

$$P = 2l + 2w \ or \ P = 2(l + w)$$

A square has four equal sides with the length s. Its length is equal to its width. The formula for the area of a square is:

$$A = s \times s$$

Finally, the area of a triangle is calculated by dividing the area of the rectangle that would be formed by the base, the altitude, and height of the triangle. Therefore, the area of a triangle is:

$$A = \frac{1}{2} \times b \times h$$

Formulas for perimeter are derived by adding length measurements of the sides of a figure. The perimeter of a rectangle is the result of adding the length of the four sides. Therefore, the formula for perimeter of a rectangle is $P = 2 \times l + 2 \times w$, and the formula for perimeter of a square is:

$$P = 4 \times s$$

The perimeter of a triangle would be the sum of the lengths of the three sides.

A triangle's perimeter is measured by adding together the three sides, so the formula is $P = a + b + c$, where $a, b,$ and c are the values of the three sides. The area is the product of one-half the base and height so the formula is:

$$A = \frac{1}{2} \times b \times h$$

It can be simplified to:

$$A = \frac{bh}{2}$$

The base is the bottom of the triangle, and the height is the distance from the base to the peak. If a problem asks to calculate the area of a triangle, it will provide the base and height.

For example, if the base of the triangle is 2 feet and the height 4 feet, then the area is 4 square feet. The following equation shows the formula used to calculate the area of the triangle:

$$A = \frac{1}{2}bh = \frac{1}{2}(2)(4) = 4 \text{ square feet}$$

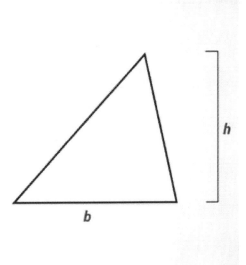

A circle's perimeter—also known as its circumference—is measured by multiplying the diameter by π.

Diameter is the straight line measured from a point on one side of the circle to a point directly across on the opposite side of the circle.

π is referred to as pi and is equal to 3.14 (with rounding).

So the formula is $\pi \times d$.

This is sometimes expressed by the formula $C = 2 \times \pi \times r$, where r is the radius of the circle. These formulas are equivalent, as the radius equals half of the diameter.

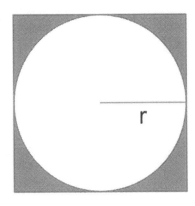

The area of a circle is calculated through the formula $A = \pi \times r^2$. The test will indicate either to leave the answer with π attached or to calculate to the nearest decimal place, which means multiplying by 3.14 for π.

Volume is a measurement of the amount of space that in a 3-dimensional figure. Volume is measured using cubic units, such as cubic inches, feet, centimeters, or kilometers.

Consider the following problem:

The total perimeter of a rectangular garden is 36 m. If the length of each side is 12 m, what is the width?

The formula for the perimeter of a rectangle is P = 2L + 2W, where P is the perimeter, L is the length, and W is the width. The first step is to substitute all of the data into the formula:

$$36 = 2(12) + 2W$$

Simplify by multiplying 2 × 12:

$$36 = 24 + 2W$$

Simplifying this further by subtracting 24 on each side, which gives:

$$36 - 24 = 24 - 24 + 2W$$

$$12 = 2W$$

Divide by 2:

$$6 = W$$

The width is 6 m. Remember to test this answer by substituting this value into the original formula:

$$36 = 2(12) + 2(6)$$

Reasoning with Shapes and Their Attributes

Understanding that Shapes in Different Categories May Share Attributes

Geometry is part of mathematics. It deals with shapes and their properties. Geometry means knowing the names and properties of shapes. It is also similar to measurement and number operations. The basis of geometry involves being able to label and describe shapes and their properties.

Flat or two-dimensional shapes include circles, triangles, hexagons, and rectangles, among others. A shape can be classified based on whether it is open like the letter U or closed like the letter O. Further classifications involve counting the number of sides and vertices (corners) on the shapes. This can help to tell the difference between shapes.

A **polygon** is a closed geometric figure in a plane (flat surface) consisting of at least 3 sides formed by line segments. These are often defined as two-dimensional shapes. Common two-dimensional shapes include circles, triangles, squares, rectangles, pentagons, and hexagons. Note that a circle is a two-dimensional shape without sides, so it is not a polygon.

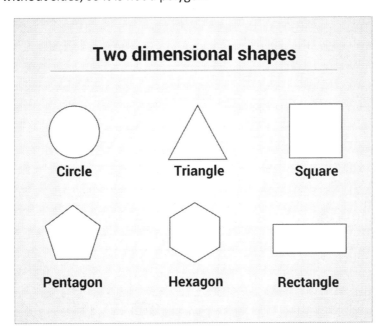

Polygons can be classified by the number of sides (also equal to the number of angles) they have. The following are the names of polygons with a given number of sides or angles:

# of Sides	Name of Polygon
3	Triangle
4	Quadrilateral
5	Pentagon
6	Hexagon
7	Septagon (or heptagon)
8	Octagon
9	Nonagon
10	Decagon

Equiangular polygons are polygons in which the measure of every interior angle is the same. The sides of equilateral polygons are always the same length. If a polygon is both equiangular and equilateral, the polygon is defined as a regular polygon.

Triangles can be further classified by their sides and angles. A triangle with its largest angle measuring 90° is a **right triangle**. A 90° angle, also called a **right angle,** is formed from two perpendicular lines. It looks like a hard corner, like that in a square. The little square draw into the angle is the symbol used to denote that that angle is indeed a right angle. Any time that symbol is used, it denotes the measure of the angle is 90°. Below is a picture of a right angle, and below that, a right triangle.

A right angle:

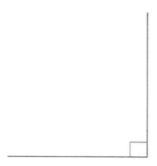

Here is a right triangle, which is a triangle that contains a right angle:

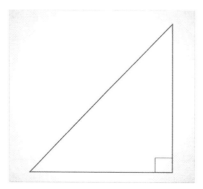

A triangle with the largest angle less than 90° is an acute triangle. A triangle with the largest angle greater than 90° is an obtuse triangle.

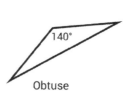

Acute
3 acute angles

Right
1 right angle

Obtuse
1 obtuse angle

A triangle consisting of two equal sides and two equal angles is an isosceles triangle. A triangle with three equal sides and three equal angles is an equilateral triangle. A triangle with no equal sides or angles is a scalene triangle.

  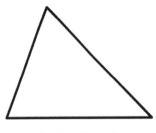

Equilateral Triangle   **Isosceles Triangle**   **Scalene Triangle**

Using Attributes to Recognize Quadrilaterals

Quadrilaterals can be further classified according to their sides and angles. A quadrilateral with exactly one pair of parallel sides is called a trapezoid. A quadrilateral that shows both pairs of opposite sides parallel is a parallelogram. Parallelograms include rhombuses, rectangles, and squares. A rhombus has four equal sides. A rectangle has four equal angles (90° each). A square has four 90° angles and four equal sides. Therefore, a square is both a rhombus and a rectangle.

Venn Diagram

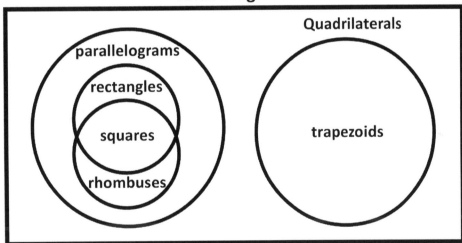

Which of the following are not rectangles?

1.   2.   3.   4.

Even though all have four sides and some are even parallel sides, numbers 3 and 4 do not fit the definition of a rectangle.

Which of the following are not parallelograms?

1. 2. 3. 4.

Even though all have four sides, numbers 1, 2, and 4 do not fit the definition of a parallelogram because they do not have two sets of parallel sides. Parallelograms need to have two pairs of parallel sides, so 1 would not fit the description.

Partitioning Shapes into Parts with Equal Areas and Expressing the Area of Each Part as a Unit Fraction of the Whole

What fraction of a pizza would one person receive if four people share the entire pizza equally? This problem involves partitioning the whole pizza into 4 smaller, same-sized parts. In order to do this, we consider the shape of a circle, which is the typical shape of a pizza. We can split up a circle into either 2, 3, or 4 parts of equal size. In this case, you would be partitioning a circle into halves, thirds, or quarters, respectively. Remember, that each part is equal in size to the other corresponding parts. Here is a picture that represents this process:

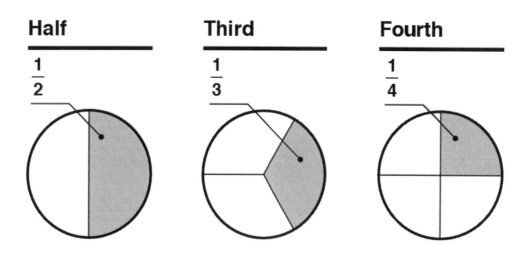

In the first circle, each part is known as "one-half," or $\frac{1}{2}$. Each segment represents $\frac{1}{2}$ of the area of the circle. In the second circle, each part is known as "one-third," or $\frac{1}{3}$, and each segment represents $\frac{1}{3}$ of the area of the circle. Finally, in the third circle, each part is known as "one-fourth," or $\frac{1}{4}$. The last picture could represent a pizza that has been cut into four equal parts. In this case, each person would get one-fourth of the pizza, and each person is actually receiving $\frac{1}{4}$ of the area of the entire pizza.

Other shapes can be partitioned into parts with equal areas. For instance, here is a triangle partitioned into 6 parts of equal area where each part represents 1/6 of the original shape.

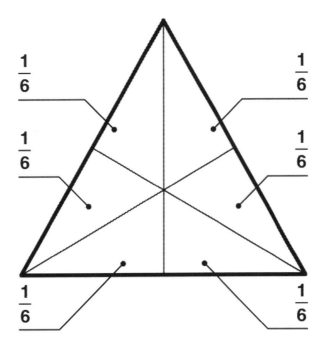

However, just because you partition a shape into a certain number of parts, it does not mean the areas have to be equal. For instance, consider the following trapezoid:

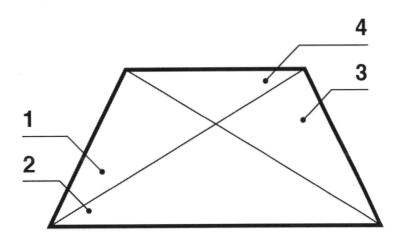

It is partitioned into 4 parts that are not equal and do not have equal areas.

Practice Test

1. How will the number 84,789 be written if rounded to the nearest hundred?
   - a. 84,800
   - b. 84,700
   - c. 84,780
   - d. 85,000

2. One digit in the following number is in **bold** and the other is underlined: 3**6**,<u>6</u>01. Which of the following statement about the underlined digit is true?
   - a. Its value is $\frac{1}{10}$ the value of the bold digit.
   - b. Its value is 10 times the value of the bold digit.
   - c. Its value is 100 times the value of the bold digit.
   - d. Its value is 60 times the value of the bold digit.

3. Angie wants to shade $\frac{3}{5}$ of this strip. Which is the correct representation of $\frac{3}{5}$?

a.

b.

c.

d.

4. Ming would like to share his collection of 16 baseball cards with his three friends. He has decided that he will divide the collection equally among himself and his friends. Which of the following shows the correct grouping of Ming's cards?

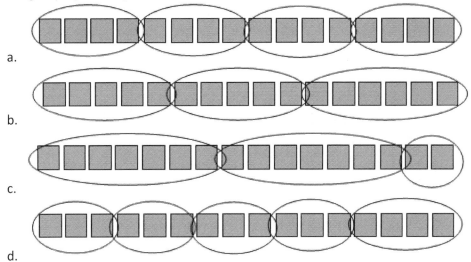

a.

b.

c.

d.

5. Which fractions are equivalent, or would fill the same portion on a number line? Select all that apply.

a. $\frac{2}{4}$ and $\frac{3}{8}$

b. $\frac{1}{2}$ and $\frac{4}{8}$

c. $\frac{3}{6}$ and $\frac{3}{5}$

d. $\frac{2}{4}$ and $\frac{5}{8}$

e. $\frac{7}{3}$ and $2\frac{1}{3}$

6. Which fraction represents the greatest part of the whole?

a. $\frac{1}{4}$

b. $\frac{1}{3}$

c. $\frac{1}{5}$

d. $\frac{1}{2}$

7. Which of the following expressions is equivalent to $\frac{8}{7}$

a. $\frac{1}{7} + \frac{1}{7} + \frac{1}{7} + \frac{1}{7} + \frac{1}{7} + \frac{1}{7} + \frac{1}{7} + \frac{1}{7}$

b. $\frac{1}{8} + \frac{1}{8} + \frac{1}{8} + \frac{1}{8} + \frac{1}{8} + \frac{1}{8} + \frac{1}{8}$

c. $\frac{1}{7} + \frac{8}{1}$

d. $\frac{7}{8} + 7$

8. Chris walks $\frac{4}{7}$ of a mile to school and Tina walks $\frac{5}{9}$ of a mile. Which student covers more distance on the walk to school?

    a. Chris, because $\frac{4}{7} > \frac{5}{9}$

    b. Chris, because $\frac{4}{7} < \frac{5}{9}$

    c. Tina, because $\frac{5}{9} > \frac{4}{7}$

    d. Tina, because $\frac{5}{9} < \frac{4}{7}$

9. A closet is filled with red, blue, and green shirts. If there are 54 shirts total, and 12 are green and 28 are red, which equation below could be used to calculate the number of blue shirts?

    a. $54 - 12 - 28 = \square$

    b. $54 = 28 - 12 + \square$

    c. $54 - \square = 28 - 12$

    d. $54 + 12 + 28 = \square$

10. Kareem arrived for his 9:00 a.m. appointment with the dentist 15 minutes early. He was taken back to see the see the dentist 30 minutes after he arrived. His cleaning took 45 minutes once he was taken back. What time was he done with the dental cleaning?

    a. 9:45 a.m.

    b. 10:00 a.m.

    c. 10:15 a.m.

    d. 10:30 a.m.

Use the following graph for Questions 11-12.

The following stem-and-leaf plot shows plant growth in cm for a group of tomato plants.

Stem	Leaf
2	0 2 3 6 8 8 9
3	2 6 7 7
4	7 9
5	4 6 9

11. What is the range of measurements for the tomato plants' growth?
    a. 29 cm
    b. 37 cm
    c. 39 cm
    d. 59 cm

12. How many plants grew more than 35 cm?
    a. 4 plants
    b. 5 plants
    c. 8 plants
    d. 9 plants

13. Which of the following is equivalent to the value of the digit 3 in the number 792.134?
    a. 3×10

    b. 3×100

    c. $\frac{3}{10}$

    d. $\frac{3}{100}$

14. These stars represent the number of ribbons Matt won at a track meet. Which equation(s) express the correct total? Select all that apply.

★★★★★★
★★★★★★
★★★★★★

    a. $3 + 6 = 9$
    b. $9 + 3 = 12$
    c. $3 \times 6 = 18$
    d. $3 \times 9 = 18$
    e. $6 \times 3 = 18$
    f. $9 \times 3 = 18$

15. Beau cut 27 lawns in one week. If he can cut 9 lawns per day, how many days of the week did Beau work?
    a. 6 days
    b. 4 days
    c. 7 days
    d. 3 days

16. Which of the following numbers is greater than (>) 220,058? Select all that apply.
    a. 220,158
    b. 221,000
    c. 202,058
    d. 220,008
    e. 217,058

17. Use the picture graph below to answer the question.

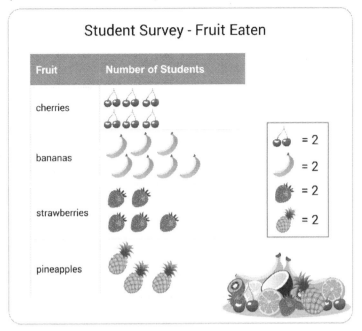

How many more bananas were eaten than pineapples?
   a. 4
   b. 5
   c. 7
   d. 8

18. The following table shows the temperature readings in Ohio during the month of January. How many more times was the temperature between 28-30 degrees than between 20-24 degrees?

Maximum Temperatures in degrees	Tally marks	Frequency
20 - 22	I	1
22 - 24	JHT II	7
24 - 26	JHT	5
26 - 28	JHT IIII	9
28 - 30	JHT JHT	10

   a. 3 times
   b. 9 times
   c. 2 times
   d. 10 times

19. A rectangle was formed out of pipe cleaner. Its length was 3 in, and its width was 8 inches. What is its area in square inches?

    a. 22 inch^2

    b. 11 inch^2

    c. 32 inch^2

    d. 24 inch^2

20. Taylor is buying things at the bake sale at his sister's basketball game. Here is the price list:

- Brownies: $0.50
- Cookies: $0.20
- Cupcakes: $0.75
- Lemon Squares: $0.60
- Milk: $0.35

He buys 1 brownie, 2 cookies, 1 cupcake, 2 lemon squares, and 1 container of milk. He gives the cashier the following money:

How much change should he receive?

    a. $0.20

    b. $0.25

    c. $0.40

    d. $0.75

21. It costs Shea $12 to produce 3 necklaces. If he can sell each necklace for $20, how much profit would he make if he sold 60 necklaces?

    a. $240

    b. $360

    c. $960

    d. $1200

22. Mom's car drove 72 miles in 1 hour. If she maintained the same speed, how far did she drive in 4 hours?

    a. 18 miles, because $72 \div 4 = 18\ miles$

    b. 16 miles, because $72 \div 3 = 24\ miles$

    c. 288 miles, because $72 \times 4 = 288\ miles$

    d. 216 miles, because $72 \times 3 = 216\ miles$

23. Which of the following are correct labels for the chart below?

Input	Calculation (Input × 3)	Output
1	1 × 3	3
2	2 × 3	6
3	3 × 3	9
4	4 × 3	12

a. Input: number of chairs; Calculation: number of chairs × number of legs on a chair; Output: number of rubber feet for chairs to order

b. Input: Number of wheels on a tricycle; Calculation: number of tricycles; Output: number of wheels in inventory

c. Input: number of tricycles; Calculation: number of wheels on a tricycle; Output: number of wheels in inventory

d. Input: number of booties for dogs; Calculation: number of dogs; Output: number of booties in inventory

24. A piggy bank contains 12 dollars' worth of nickels. A nickel weighs 5 grams, and the empty piggy bank weighs 1050 grams. What is the total weight of the full piggy bank?
   a. 1,110 grams
   b. 1,200 grams
   c. 2,250 grams
   d. 2,200 grams

25. What is $\frac{420}{100}$ rounded to the nearest whole number?
   a. 4
   b. 3
   c. 5
   d. 6

26. A construction company is building a new housing development with the property of each house measuring 30 feet wide. If the length of the street is zoned off at 345 feet, how many houses can be built on the street?
   a. 11
   b. 115
   c. 11.5
   d. 12

27. The total perimeter of a rectangle is 36 cm. If the length is 12 cm, what is the width?
    a. 3 cm
    b. 12 cm
    c. 6 cm
    d. 8 cm

28. The following questions are based on this graph of test scores for three students who have classes with four teachers:

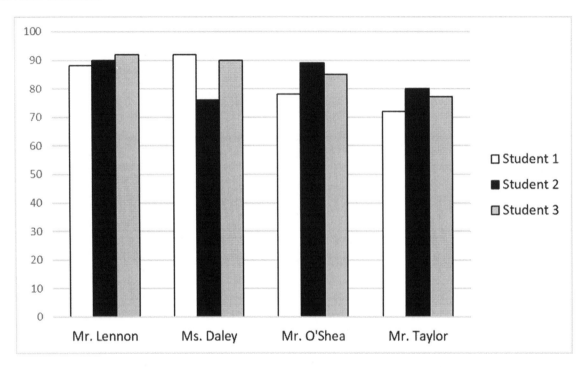

Based on the graph, how many more points did student 3 get on her test in Mr. O'Shea's class than on her test in Mr. Taylor's class?
    a. 5
    b. 8
    c. 10
    d. 14

29. Bernard can make $80 per day. If he needs to make $300 and only works full days, how many days will this take? Enter your answer in the box.

The table below shows the number of students in Ms. Jackson' class who play each sport.

Sports Played By Students in Ms. Jackson's Class

Sport	Frequency
Soccer	ⵏⵏ ‖
Swimming	‖
Track	‖‖
Baseball	ⵏⵏ ‖
Basketball	ⵏⵏ ‖
Tennis	‖

30. Which of the following dot plots correctly represents the data in the table?

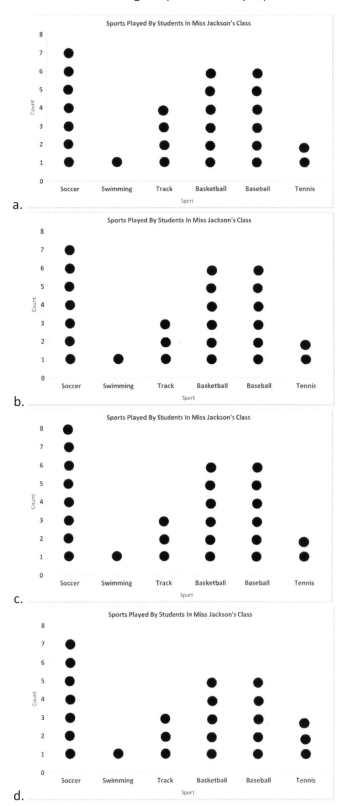

31. Jessica buys 10 cans of paint. Red paint costs $1 per can and blue paint costs $2 per can. In total, she spends $16. How many red cans did she buy?

    a. 2

    b. 3

    c. 4

    d. 5

32. The perimeter of a 6-sided polygon is 56 cm. The length of three of the sides are 9 cm each. The length of two other sides are 8 cm each. What is the length of the missing side in cm? Enter your answer in the box.

33. If Amanda can eat two times as many mini cupcakes as Marty, what would the missing values be for the following input-output table?

Input (number of cupcakes eaten by Marty)	Output (number of cupcakes eaten by Amanda)
1	2
3	
5	10
7	
9	18

    a. 6, 10

    b. 3, 11

    c. 6, 14

    d. 4, 12

34. Which of the following is not a parallelogram?

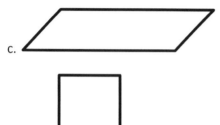

a.

b.

c.

d.

35. Which of the following figures is not a polygon?
    a. Decagon
    b. Cone
    c. Triangle
    d. Rhombus

36. A teacher cuts a pie into 6 equal pieces and takes five away. What fraction of the pie remains?
    a. $\frac{1}{6}$

    b. $\frac{1}{5}$

    c. $\frac{5}{6}$

    d. $\frac{6}{5}$

37. Which is the correct decomposition of the number 36,901?
    a. 36,000 + 901
    b. 3,000 + 600 + 90 + 1
    c. 30,000 + 6,000 + 900 + 1
    d. 30,000 + 6,000 + 900 + 10

38. Which array would help determine the missing number in the following equation?

$$24 \div ? = 8$$

a.

$$24 \div 2 = 8$$

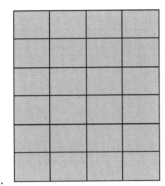

b.

$$24 \div 6 = 4$$

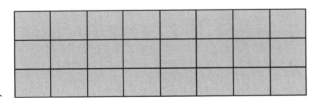

c.

$$24 \div 3 = 8$$

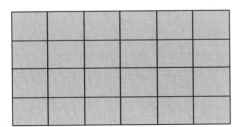

d.

$$24 \div 4 = 8$$

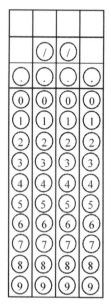

39. Mandy wants to re-carpet her weight room with carpet squares, and needs to figure out how many to buy. To do this, Mandy must calculate the total area of her irregularly shaped weight room. Using the following figure, what is the total area in square feet? (Each square pictured is 1 ft × 1 ft.) Enter your answer in the box.

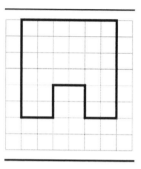

40. Marcus wants to paint a part of a wall of his bedroom orange. He has only enough paint to cover an area of 6 sq ft. If the total area of the wall is 36 sq ft, which of the following could he NOT cover with his 6 sq ft allotment of paint?

a.

b.

c.

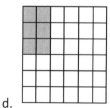
d.

41. If Mae needs to be at her appointment by 3:45 p.m., which three of the following activities (listed by length of time) could she complete, beginning at 3:05 p.m.? (Use the clock face to determine the answer.)

Walk a lap in the park = 30 minutes

Buy a soda at the store next to her appointment = 10 minutes

Call her father = 15 minutes

Play a game on her phone = 15 minutes

Watch a short video on her phone = 30 minutes

    a. Buy a soda at the store next to her appointment, call her father, and play a game on her phone.
    b. Watch a short video on her phone, buy a soda at the store next to her appointment, and play a game on her phone.
    c. Walk a lap in the park, call her father, and watch a short video on her phone.
    d. Watch a short video on her phone, buy a soda at the store next to her appointment, and call her father.

42. Which of the following is the appropriate tool for measuring the amount of water in a bathtub full enough to cover an adult?
    a. Measuring cup
    b. Tablespoon
    c. Liter container
    d. Gallon container

43. How much money (in dollars) is pictured?

   a. $1.66
   b. $2.66
   c. $1.56
   d. $2.16

44. How many more votes do blueberries have than bananas on the graph of a Favorite Fruit survey?

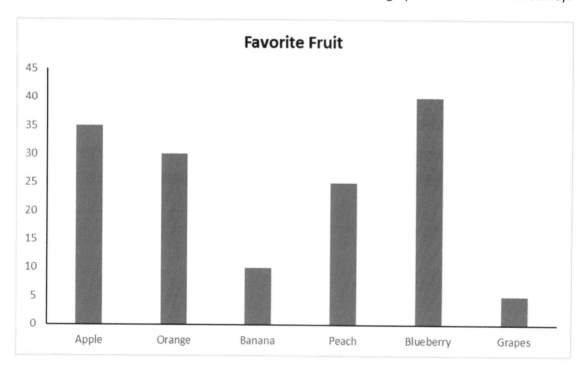

   a. 10
   b. 15
   c. 20
   d. 30

45. Keith started tiling his bathroom floor with the help of his dad. They are using square tiles that each have a side length of 1 foot.

They are going to continue tiling the floor so that it is completely covered in the square tiles with no overlaps. What is the area of the floor?

    a. $110 \; ft$
    b. $112 \; ft$
    c. $100 \; ft^2$
    d. $112 \; ft^2$

46. Shotaro is a champion swimmer and has a big collection of ribbons from swim meets that he displays above his bed. The top row contains 16 blue ribbons. Below that are 3 rows that each contain 13 red ribbons. The last four rows each contain 9 yellow ribbons. How many ribbons does he have in total? Enter your answer in the box.

47. What is the value, to the nearest tenths place, of the point indicated on the following number line? Enter your answer in the box.

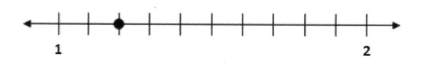

48. Which of the following best represents the length of the paperclip?

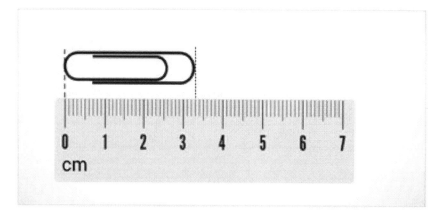

a. $3\frac{1}{2}$ cm

b. $3\frac{3}{10}$ cm

c. $4\frac{1}{2}$ cm

d. $4\frac{7}{10}$ cm

Answer Explanations

1. A: 84,800. The hundred place value is located three digits to the left of the decimal point (the digit 7). The digit to the right of the place value is examined to decide whether to round up or keep the digit. In this case, the digit 8 is 5 or greater so the hundred place is rounded up. When rounding up, any digits to the left of the place value being rounded remain the same and any to its right are replaced with zeros. Therefore, the number is rounded to 84,800.

2. B: The underlined digit is the 6 in 6,000. The bold digit is the 6 in 600. Because 6,000 is equal to 6000×10, we know that the underlined 6 has a value that is 10 times that of the bold 6. Additionally, the base-10 system we use helps us determine that the place value increases by a multiple of ten when you go from the right to the left.

3. D: This solution shows the strip separated into 5 pieces, which is necessary for it to be filled in to show $\frac{3}{5}$. Choice A shows the strip filled to $\frac{1}{2}$, and Choice B shows the strip filled to $\frac{2}{4}$, which is also $\frac{1}{2}$. Neither of these selections is correct. While Choice C shows 3 portions filled, the total number of portions is only 4, making the fraction filled $\frac{3}{4}$. This is also an incorrect choice.

4. A: This choice shows that Ming plus his three friends (1 + 3 = 4) is the number of divisions necessary to split the lot of cards evenly (16 ÷ 4 = 4). There would need to be four groups of 4 cards each, or $\frac{4}{16}$, which is $\frac{1}{4}$ of the total cards. The other choices do not correctly divide the cards into even groupings.

5. B & E: $\frac{1}{2}$ is the same fraction as $\frac{4}{8}$, and would both fill up the same portion of a number line.

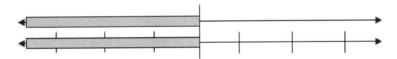

The other answer choice that contains equivalent fractions is E, $\frac{7}{3}$ and $2\frac{1}{3}$. $\frac{7}{3}$ is the improper fraction for the mixed number $2\frac{1}{3}$. Recall that to convert a mixed number to an improper fraction, multiply the denominator by the whole number then add the numerator and place this sum over the denominator (as your new numerator). In this case,

$$2\frac{1}{3} \rightarrow 3 \times 2 + 1 \rightarrow \frac{7}{3}.$$

None of the other choices represent equivalent portions to each other, as seen below.

A.

C.

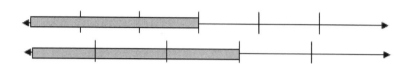

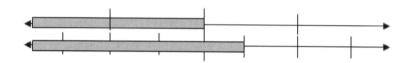

D.

6. D: Even though all of the fractions have the same numerator, this is the one that represents the greatest part of the whole. All other choices are smaller portions of the whole, as seen by this graphic representation.

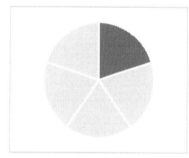

$\frac{1}{4}$

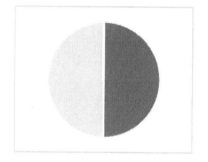

$\frac{1}{3}$

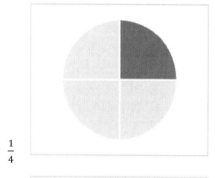

$\frac{1}{5}$

$\frac{1}{2}$

7: A: $\frac{8}{7}$ is the same as $8 \times \frac{1}{7}$, which is represented by the first option:

$$\frac{1}{7} + \frac{1}{7} + \frac{1}{7} + \frac{1}{7} + \frac{1}{7} + \frac{1}{7} + \frac{1}{7} + \frac{1}{7}$$

This can be thought of as cutting a pie into seven slices and then serving 8 slices. Since you need more slices than you have in your pie, you actually need to cut up two pies and take one piece from the second pie. This is because $\frac{8}{7}$ is an improper fraction, which means the numerator (top number) is greater than the denominator (bottom number).

8. A: In order to compare the fractions $\frac{4}{7}$ and $\frac{5}{9}$, a common denominator must be used. The least common denominator is 63, which is found by multiplying the two denominators together (7×9). The conversions are as follows:

$$\frac{4}{7} \times \frac{9}{9} = \frac{36}{63}$$

$$\frac{5}{9} \times \frac{7}{7} = \frac{35}{63}$$

Although they walk nearly the same distance, $\frac{4}{7}$ is slightly more than $\frac{5}{9}$ because $\frac{36}{63} > \frac{35}{63}$. Remember, the sign > means "is greater than." Therefore, Chris walks further than Tina, and Choice *A* correctly shows this expression in mathematical terms.

9. A: The sum of the shirts is 54. So, we know that the number of green shirts, plus the number of red shirts, plus the number of blue shirts is 54. By subtracting the number of green shirts (12) and the number of red shirts (28) from the total number of shirts (54), we can calculate the number of blue shirts, which make up the remaining portion of the total.

10. B: Kareem arrived to his appointment at 8:45 a.m., since that's 15 minutes before 9:00 a.m., which was his scheduled time. He was taken back at 9:15 a.m., since 30 minutes after 8:45 a.m. is 9:15 a.m. His cleaning was 45 minutes, so he was done at 10:00 a.m.

11. C: The range of the entire stem-and-leaf plot is found by subtracting the lowest value from the highest value, as follows: $59 - 20 = 39$ cm. All other choices are miscalculations read from the chart.

12. C: To calculate the total greater than 35, the number of measurements above 35 must be totaled; 36, 37, 37, 47, 49, 54, 56, 59 = 8 measurements. Choice *A* is the number of measurements in the 3 categories, Choice *B* is the number in the 4 and 5 categories, and Choice *D* is the number in the 3, 4, and 5 categories.

13. D: $\frac{3}{100}$. Each digit to the left of the decimal point represents a higher multiple of 10 and each digit to the right of the decimal point represents a quotient of a higher multiple of 10 for the divisor. The first digit to the right of the decimal point is equal to the value $\div 10$. The second digit to the right of the decimal point is equal to the value $\div (10 \times 10)$, or the value $\div 100$.

14. C & E: There are 3 rows and 6 columns of stars, representing a total of 18 stars, modeled by the equation $3 \times 6 = 18$ ir $6 \times 3 = 18$, which are equal due to the commutative property. Choices *A* and *B* do not represent the multiplication necessary to compute the total number of stars. Choices *D* and *F* do not represent a correct multiplication equation. $3 \times 9 = 27$, not 18.

15. D: The equation to model this is to divide 9 into the total number of lawns mowed.

$$27 \div 9 = 3$$

A quick check of multiplying 9 × 3 gives the original total of 27 lawns. Choices *A*, *B*, and *C* all represent miscalculations of the division necessary to calculate this answer.

16. A & B: These choices can be determined by comparing the place values, beginning with that which is the farthest left; hundred-thousands, then ten-thousands, then thousands, then hundreds. It is in the hundreds place that Choice *A* is larger. It is in the thousands place that Choice *B* is larger. Choice *C* is smaller in the ten-thousands place, Choice *D* is smaller in the tens place, and Choice *E* is smaller in the ten-thousands place.

17. D: Using the key, it can be seen that each fruit symbol is equivalent to 2 counts of that fruit. There are 7 bananas pictured, which means 14 bananas were eaten because 7 × 2 = 14. There were 6 pineapples eaten because 3 are pictured and 3 × 2 = 6. Then, the difference must be found: 14 − 6 = 8, so 8 more bananas were eaten. Therefore, Choice *D* is the correct answer.

18. C: To calculate this, the following equation is used: $10 - (7 + 1) = 2$. The number of times the temperature was between 28-30 degrees was 10. Finding the total number of times the temperature was between 20-24 degrees requires totaling the categories of 20-22 degrees and 22-24 degrees, which is $7 + 1 = 8$. This total is then subtracted from the other category in order to find the difference. Choice *A* only subtracts the 28-30 degrees from the 22-24 degrees category. Choice *B* only subtracts the 28-30 degrees category from the 20-22 degrees category. Choice *D* is simply the number from the 28-30 degrees category.

19. D: Area = length × width. Therefore, the area of the rectangle is equal to $3 \text{ in} \times 8 \text{ in} = 24 \text{ in}^2$.

20. C: First, we need to add up the money Taylor had: 2 $1 bills, 5 quarters, 3 dimes, and 1 nickel:

$$(2 \times \$1.00) + (5 \times \$0.25) + (3 \times \$0.10) + \$0.05 = \$3.60$$

Taylor buys 1 brownie, 2 cookies, 1 cupcake, 2 lemon squares, and 1 container of milk. Using the price list, we can write an equation to represent the total cost of his purchases:

$$\$0.50 + (2 \times \$0.20) + \$0.75 + (2 \times \$0.60) + \$0.35 = \$3.20$$

To calculate this change, we subtract the total cost of his purchases ($3.20) from what he gave the cashier ($3.60) to get $0.40.

21. C: In order to calculate the profit, we need to create an equation that models the total income minus the cost of the materials.

$60 × 20 = $1,200 total income
60 ÷ 3 = 20 sets of materials
20 × $12 = $240 cost of materials
$1,200 − $240 = $960 profit

Choice *A* is not correct, as it is only the cost of materials. Choice *B* is not correct, as it is a miscalculation. Choice *D* is not correct, as it is the total income from the sale of the necklaces.

22. C: Mom can drive 72 miles each hour and she's driving for four hours, so we can recognize this as a multiplication problem that can be represented by:

$$72 \times 4 = 288 \text{ miles}$$

23. C: These labels correctly describe a real-world application of the input-output table shown. The number of tricycles would need to be multiplied by 3 (the number of wheels on a tricycle) in order to find the number of total wheels in a store's inventory. Choice *A* is not a correct modeling of a real-world situation. A stable chair would have 4 legs, not 3. Choice *B* is incorrect as it mixes up the number of wheels on a tricycle with the number of tricycles. The number of wheels cannot be the variable (changing) item for this calculation. Choice *D* does something similar as Choice *B*, by mixing up the variable and the multiplier; dogs would have a set number of paws, not one that would change.

24. C: A dollar contains 20 nickels. Therefore, if there are 12 dollars' worth of nickels, there are $12 \times 20 = 240$ nickels. Each nickel weighs 5 grams. Therefore, the weight of the nickels is $240 \times 5 = 1,200$ grams. Adding in the weight of the empty piggy bank, the filled bank weighs 2,250 grams.

25. A: Dividing by 100 involves mean shifting the decimal point of the numerator to the left by 2. The result is 4.2 and rounds to 4.

26. A: 11. To determine the number of houses that can fit on the street, the length of the street is divided by the width of each house: $345 \div 30 = 11.5$. Although the mathematical calculation of 11.5 is correct, this answer is not reasonable. Half of a house cannot be built, so the company will need to either build 11 or 12 houses. Since the width of 12 houses (360 feet) will extend past the length of the street, only 11 houses can be built.

27. C: The formula for the perimeter of a rectangle is P=2l+2w, where P is the perimeter, l is the length, and w is the width. The first step is to substitute all of the data into the formula:

$$36 = 2(12) + 2W$$

Simplify by multiplying 2x12:

$$36 = 24 + 2W$$

Simplifying this further by subtracting 24 on each side, which gives:

$$36 - 24 = 24 - 24 + 2W$$

$$12 = 2W$$

Divide by 2:

$$6 = W$$

The width is 6 cm. Remember to test this answer by substituting this value into the original formula:

$$36 = 2(12) + 2(6)$$

28. B: To calculate the difference between the two scores for Student 3, subtract the score from Mr. Taylor's class (77) from the score in Mr. O'Shea's class (85):

$$85 - 77 = 8$$

			4

(grid-in answer: 4)

29. The number of days can be found by taking the total amount Bernard needs to make and dividing it by the amount he earns per day:

$$\frac{300}{80} = \frac{30}{8} = \frac{15}{4} = 3.75$$

But Bernard is only working full days, so he will need to work 4 days, since 3 days is not a sufficient amount of time.

30. B: The dot plot in Choice *B* is correct because, like the table, it shows that 7 students play soccer, 1 swims, 3 run track, 6 play basketball, 6 play baseball, and 2 play tennis. Each dot represents one student, just like one hash mark in the table represents one student.

31. C: We are trying to find x, the number of red cans. The equation can be set up like this:

$$x + 2(10 - x) = 16$$

The left x is actually multiplied by $1, the price per red can. Since we know Jessica bought 10 total cans, $10 - x$ is the number blue cans that she bought. We multiply the number of blue cans by $2, the price per blue can.

That should all equal $16, the total amount of money that Jessica spent. Working that out gives us:

$$x + 20 - 2x = 16$$

$$20 - x = 16$$

$$x = 4$$

32. Perimeter is found by calculating the sum of all sides of the polygon. $9 + 9 + 9 + 8 + 8 + s = 56$, where s is the missing side length. Therefore, 43 plus the missing side length is equal to 56. The missing side length is 13 cm.

33. C: The situation can be described by the equation $? \times 2$. Filling in for the missing numbers would result in $3 \times 2 = 6$ and $7 \times 2 = 14$. Therefore, the missing numbers are 6 and 14. The other choices are miscalculations or misidentification of the pattern formed by the table.

34. A: A parallelogram has two sets of parallel sides. Choice *A* is a trapezoid and only has one set of parallel sides. The rest of the answer choices have two sets.

35. B: Cone. A polygon is a closed two-dimensional figure consisting of three or more sides. A decagon is a polygon with 10 sides. A triangle is a polygon with three sides. A rhombus is a polygon with 4 sides. A cone is a three-dimensional figure and is classified as a solid.

36. A: If a pie was cut into 6 pieces, each piece would represent $\frac{1}{6}$ of the pie. If five pieces were taken away, $\frac{5}{6}$ would be removed and just $\frac{1}{6}$ of the pie would be left over because:

$$\frac{6}{6} - \frac{5}{6} = \frac{1}{6}$$

37. C: This answer has the proper place values for each of the digits: 3 is in the ten-thousands place, 6 is in the thousands place, 9 is in the hundreds place, and 1 is in the ones place. Choice *A* is not correct because it does not decompose the entire number into all of its place values. Choice *B* is not correct because the 3, 6, and 9 are decomposed into the wrong place values. Choice *D* is not correct because the 1 is in the tens place, not the ones place.

38. C: The array displayed in this solution is the only one that correctly represents the total number of items (24) evenly divided into 8 columns of 3 items each. Choice *A* divides the total by 2, resulting in an incorrect solution of 12. Choice *B* divides the total by 4, resulting in an incorrect solution of 6. Choice *D* divides the total into 6 columns of 4, which does not solve the initial problem of dividing 24 by 8.

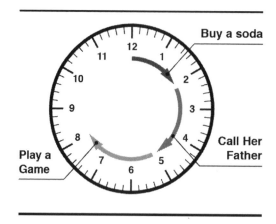

39. The solution could be calculated by dividing the shape up as follows:

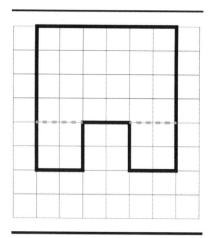

The area is calculated as the sum of the three rectangles:

$$6 \times 4 = 24$$
$$2 \times 2 = 4$$
$$2 \times 2 = 4$$

$$24 + 4 + 4 = 32 \text{ sq ft}$$

40. B: The choice does not show a portion covering 6 sq ft. It shows a coverage area of $3 \times 3 = 9$ sq ft. Choice *A* shows a coverage area of $1 \times 6 = 6$ sq ft, Choice *C* shows a coverage area of $2 \times 3 = 6$ sq ft, and Choice *D* shows a coverage area of $3 \times 2 = 6$ sq ft. Choices *A*, *C*, and *D* all show the same area, in differing shapes.

41. A: This combination of three items is the only one to total 40 minutes, as seen on the following clock face.

Choice *B* totals 55 minutes, Choice *C* totals 75 minutes, and Choice *D* totals 55 minutes.

42. D: The gallon container would measure the largest capacity of water and therefore be the correct choice. Choices *A* and *B* would take a long time to measure the amount of water contained in a bathtub, and Choice *C*, while slightly faster, is still an impractical selection.

43. B: The total of money is as follows:

$$\$1.00 + \$1.00 + \$0.25 + \$0.25 + \$0.10 + \$0.05 + \$0.01 = \$2.66$$

44. D: The total number of blueberry votes on the graph is 40, and the total number of banana votes is 10. To find the difference, solve the following equation: $40 - 10 = 30$. The difference in votes is 30. All of the other selections involve misreading or miscalculating the number of votes.

45. D: There are 28 tiles along the length of the rectangular floor and 4 tiles for the width. Each square tile has a length of 1 foot so that means the floor is 28 feet by 4 feet. To find the area, multiply 28 feet × 4 feet, which equals 112 square feet or $112 \, ft^2$. Remember that area is measured in square units.

46. The sum of Shotaro's ribbons equals the total number of blue, red, and yellow ribbons. He has 16 blue ribbons since there is one row of 16 blue ribbons. He has 39 red ribbons because there are 3 rows that each contain 13 red ribbons and 3 x 13 = 39. There are 36 yellow ribbons because there are 4 rows that each contain 9 yellow ribbons. The total is:

$$16 + 39 + 36 = 91 \; ribbons$$

47. The number line is divided into 10 sections, so each portion represents 0.1. Because the number line begins at 1 and ends at 2, the number in question would be between those two numbers. Since there are only two portions out of ten marked, this represents the number 1.2.

48. B: The paperclip is $3\frac{3}{10}$ cm. One end is properly lined up at 0 cm, and then the far end falls between 3 and 4 centimeters, so we can rule out Choices *C* and *D,* since they are greater than 4 cm. Looking more carefully, it can be seen that the longer tick mark between each numbered centimeter represents ½ of a centimeter. The paperclip doesn't reach the $3\frac{1}{2}$ cm mark. It only extends to 3 little tick marks past the 3 cm line, so it is $3\frac{3}{10}$ cm. Each little tick mark represents $\frac{1}{10}$ cm.

Dear Customer,

Thank you for purchasing our FSA Grade 3 study guide. We hope that we exceeded your expectations.

Our goal in creating this guide was to cover all of the topics that are likely to appear on the test. We also strove to make our practice questions as similar as possible to what will be seen on test day. With that being said, if you found something that you feel was not up to your standards, please send us an email and let us know.

We have study guides in a wide variety of fields. If you're interested in one, try searching for it on Amazon or send us an email.

Thanks Again and Happy Testing!
Product Development Team
info@studyguideteam.com

FREE Test Taking Tips DVD Offer

To help us better serve you, we have developed a Test Taking Tips DVD that we would like to give you for FREE. **This DVD covers world-class test taking tips that you can use to be even more successful when you are taking your test.**

All that we ask is that you email us your feedback about your study guide. Please let us know what you thought about it – whether that is good, bad or indifferent.

To get your **FREE Test Taking Tips DVD**, email freedvd@studyguideteam.com with "FREE DVD" in the subject line and the following information in the body of the email:

     a. The title of your study guide.

     b. Your product rating on a scale of 1-5, with 5 being the highest rating.

     c. Your feedback about the study guide. What did you think of it?

     d. Your full name and shipping address to send your free DVD.

If you have any questions or concerns, please don't hesitate to contact us at freedvd@studyguideteam.com.

Thanks again!

Made in the USA
Columbia, SC
10 March 2022

57511004R00052